TRADITIONAL and INCENTIVE REGULATION

TRADITIONAL and INCENTIVE REGULATION

Applications to Natural Gas Pipelines in Canada

Robert L. Mansell
Jeffrey R. Church

Van Horne Institute
for International Transportation
and Regulatory Affairs

Canadian Cataloguing in Publication Data
Mansell, Robert L.
Traditional and incentive regulation

Includes bibliographical references.
ISBN 0-88953-209-5

1. Natural gas pipelines–Canada. 2. Gas industry–Canada–Government policy. 3. Gas industry–Canada–Economic aspects. 4. Natural gas pipelines–Canada–Finance.
I. Church, Jeffrey, 1962- II. Van Horne Institute for International Transportation and Regulatory Affairs. III. Title.
HD9581.C32M36 1995 388.5′6′0971 C95-910831-9

Printed and bound in Canada.

This book is printed on acid-free paper.

THE VAN HORNE INSTITUTE

The Van Horne Institute for International Transportation and Regulatory Affairs is a not-for-profit organization, established in 1992, and affiliated with The University of Calgary. The Institute has been established to address important transportation and related regulatory issues such as legislation, taxes, subsidies, technology and economics, and is funded through memberships and fees for services.

The Institute has developed a five-year business plan, which includes the following major objectives:

- To contribute to the expansion at the post-secondary level of **education and training**, through programs which will provide both the private and public sectors with more productive labour forces and effective management.
- To undertake Institute or client-sponsored **research** projects including, but not limited to, matters of an economic, public policy, managerial, legal, environmental, or technical nature.
- To provide timely **forums** (workshops, conferences, seminars, speakers) which will influence public and private sector policy-makers on critical transportation matters.
- To **publish** and disseminate results of forums, non-confidential research and presently unpublished works which will assist decision-makers in resolving issues.

To Tina and Justin
RLM

To Maureen, Richard and Elizabeth
JRC

Contents

Figures

Preface

This study was initiated in early 1993 by Dr. Robert L. Mansell and Dr. Jeffrey R. Church of the Department of Economics, The University of Calgary. The primary objective was to examine alternative regimes for the regulation of major natural gas pipelines in Canada, including clarifying and analyzing key issues and advancing the debate relating to these regulatory alternatives.

Publication of this study by The Van Horne Institute is consistent with one of the Institute's key objectives; namely, to publish and disseminate the results of non-confidential research that can be of assistance to decision-makers in resolving issues. However, the analysis and conclusions contained in the study are those of the authors and not necessarily those of the Institute.

The Institute would like to express its gratitude to a number of individuals in the natural gas transmission and distribution industry, regulatory lawyers and others who served as referees in reviewing an earlier draft of the study. Their comments were most helpful to the authors in the preparation of the final manuscript.

The Institute would also like to express its thanks to Drs. Mansell and Church for their efforts in undertaking this comprehensive study, and for the many hours which they devoted to the basic research on the topic and to the preparation of the finished study.

THE VAN HORNE INSTITUTE
August, 1995

Acknowledgements

This book is the outcome of our research and teaching in the areas of regulatory economics and industrial organization, along with our involvement in various forums over a number of years dealing with various aspects of the natural gas sector. We have benefited greatly from discussions with many individuals in the oil and gas industry, the pipeline industry and regulatory agencies. The Canadian Petroleum Association, Foothills Pipe Lines Ltd. and NOVA Gas Transmission Ltd. provided financial support for some of the earlier research on the regulation of gas transmission in Canada. However, the views, opinions and conclusions expressed in this study are entirely ours and should not be interpreted as reflecting the views or positions of these organizations or The Van Horne Institute.

We would like to express our appreciation for the able research assistance provided by George Given, Jody Proctor, and Michael Vaney, particularly for his contributions to the institutional survey on alternative incentive mechanisms in Section 6.1. We are especially grateful to Andrew Bradford for help in preparing the final draft, to the anonymous referees who provided many valuable comments, and to Gordon Pearce and The Van Horne Institute for assistance in bringing this study to fruition.

1 Introduction

1.1 BACKGROUND

There have been dramatic shifts in the structure and operation of the Canadian natural gas industry since the mid-1980s and further changes are inevitable. With the deregulation of gas markets and prices, the roles of the main players in the production, transportation, and distribution of gas have been significantly altered. Especially important has been the elimination of the merchant function[1] for long-distance pipelines, the opening of access to transmission facilities, and the encouragement of competitive markets for natural gas. The completion of similar reforms in the United States with FERC Order 636 and the adjustments as gas markets in both the United States and Canada have moved from overall surpluses to a more balanced situation will undoubtedly result in pressure for change in the roles of these players and in the regulatory environment.

Equally important have been the changes in gas prices, production, and the intensity of competition for markets. While fieldgate prices have been somewhat higher in recent years, by 1992 they had fallen to levels that, in real terms, were about one-third of those a decade earlier. With much smaller declines in transmission (and distribution) costs, the percentage of delivered prices accounted for by transmission (and distribution) increased markedly. Also, as a result of market and price deregulation and the low prices for natural gas, demand, particularly in export markets, grew rapidly. The construction of new transportation facilities to serve these demands has put upward pressure on tolls for many transmission systems. These developments have coincided with considerably more intense competition in domestic and export markets.

The combination of these events has made it clear that greater efficiency in all aspects of gas supply is critical and it has generally sharpened the focus of regulators on efficiency issues. Many gas

---

1 This refers to the provision of bundled gas transportation and gas commodity services.

producers, having substantially reduced their costs as prices fell, have argued that significant efficiency gains and cost reductions would also be possible for the transportation component with the appropriate incentives or regulatory changes.[2] This environment has created many new challenges for both the natural gas industry and those who regulate it. While market forces have generally been substituted for direct regulation of the gas commodity trade, the gas transmission (and distribution) components remain highly regulated. A major challenge is to ensure that the regulation of these components is consistent with the operation of properly-functioning, competitive gas markets and with the overall objective of greater efficiency.

Even limiting the focus to the major gas transmission systems, as is done in this study, the issues, challenges, and alternatives remain numerous and complex. At the most fundamental level there are questions concerning the very rationale for regulation. For example, arguments have been advanced that competition in many markets by gas via alternative pipeline routes, by gas from other supply basins, by other types of energy, or by the threat of entry by new pipelines may be sufficient to curtail the market power of particular pipelines, making traditional regulation unnecessary. More common are the proposals which accept the need for regulation but involve the introduction of various incentive mechanisms to encourage greater efficiency or the streamlining of regulatory processes.

Examples of other issues include: the appropriateness of greater tolling flexibility and the introduction of competitive capacity allocation schemes to optimize the use of existing capacity and to provide proper market signals; the determination of optimal reliability levels; the regulation of capacity expansions and the appropriate tolling methodology (for example, rolled-in versus incremental tolls) for new capacity; and optimal depreciation rules and the advantages and disadvantages of various toll structures (for example, traditional front-end loaded versus levelized tolls).[3]

These challenges and issues are unquestionably intricate and difficult. In part this is due to the extreme complexity of the natural gas industry. It involves huge capital outlays at every stage and specialized, long-lived assets which cannot be used for other purposes; the lead times are often very lengthy; there is considerable uncertainty with respect to gas supply and there are often large variations in demands and prices; it requires high levels of coordination from the fieldgate

2 For example, see Canadian Association of Petroleum Producers (1992).
3 For a discussion of these and other issues in a U.S. context, see Lyon (1990).

along many different systems to the end-user; and it involves a myriad of complex contracts. Deregulation has added further intricacy. Previously, the merchant pipelines had the capability and incentives to plan and coordinate all aspects of the supply of the commodity and transportation. However, under deregulation these activities have become separated and reallocated among the various parties involved in the production, transportation and distribution / end-use of natural gas. Since the 'invisible hand' of the market operates on only a subset of these activities, there remains considerable potential, in the absence of enlightened regulation, for dysfunction, dislocations and inefficiencies.

Additional complexities arise from the continuing evolution of major pipelines in the U.S. and Canada into a network of integrated systems. This means that many issues and their effective resolution should ideally be addressed on a system-wide basis. However, this is often impossible since these systems frequently cross numerous boundaries of independent regulatory agencies and there is no mechanism to ensure regulatory consistency across these jurisdictions.

1.2 OBJECTIVES

There have been a number of developments that provide assistance in the analysis and resolution of these complex issues. Various hearings and workshops, particularly by the National Energy Board, have generated detailed examinations of such topics as tolling methodology and incentive regulation.[4]

Over the last decade there has also been considerable experience and knowledge accumulated from regulatory changes and experiments with respect to other utilities involving such activities as telecommunications, airlines and electricity, as well as those providing pipeline service in other jurisdictions. In addition, there has been a rapid growth in the literature on regulation, especially since the mid-1980s. This has provided new tools that can be used in the analysis of the regulatory issues related to gas pipeline transmission, and also includes the results of considerable research on all aspects of regulation.

The general objective in this study is to bring together and apply these developments in an examination of some of the main alternatives concerning the regulation of the major gas pipelines in Canada. The intent is, first and foremost, to provide a framework that will assist in

---

4 On these two topics see the submissions and Reasons For Decision for NEB GH-5-89 (National Energy Board 1990a) and the submissions and Summary of Discussion for the NEB Incentive Regulation Workshop (National Energy Board 1993a).

clarifying and analyzing the key issues. It is also hoped that the concepts, techniques and research results will resolve some of the important questions and, at the very least, advance the debates on the alternatives.

The specific topics addressed include: the implications of gas market and price deregulation for pipelines; the rationale for the regulation of gas transmission systems; and the relative merits of traditional cost of service and various incentive regulation schemes.

While there are dozens of gas pipelines operating in Canada,[5] this study focuses on those lines which account for the bulk of the gas flows to Canadian and export markets. These include the systems of Alberta Natural Gas Company Ltd. (ANG), Foothills Pipe Lines Ltd. (Foothills), NOVA Corporation of Alberta (NOVA), Trans Quebec and Maritimes Pipeline Inc. (TQM), TransCanada PipeLines Limited (TCPL), and Westcoast Energy Inc (Westcoast). Except for the NOVA system, these are regulated by the National Energy Board (NEB). The NOVA system was traditionally regulated by both the Alberta Energy Resources Conservation Board (ERCB) and the Alberta Public Utilities Board (APUB). As of January 1, 1995 the ERCB and the APUB were combined and this new regulatory body, the Alberta Energy and Utilities Board or AEUB, has regulatory authority over NOVA Gas Transmission Limited (NGTL).

These pipeline companies all operate major, long-distance transmission systems. NGTL also provides some gathering services within Alberta, and the Westcoast system also includes gas processing and gathering facilities in British Columbia. These pipelines are referred to in the study as the major or long-distance pipelines.

1.3 OUTLINE

Various aspects of the natural gas industry and the regulatory system that are relevant to the subsequent analyses are outlined in Section 2. This includes an overview of the development and structure of the gas pipeline industry and a discussion of the main changes in the market and regulatory environment which have implications for the regulation of pipelines. The nature of the demands for regulatory change are also indicated.

Section 3 focuses on basic regulatory concepts and objectives. It outlines the main theories of regulation, the rationale for regulating

5 For example, as of 1992 there were 30 operating pipelines regulated by the National Energy Board alone.

natural gas pipelines and the rules for optimal pricing. The standard objectives and criteria used in the evaluation of regulatory methods and regimes are also outlined.

Section 4 deals with traditional cost of service (COS) regulation. It begins with a description of the key elements of COS regulation and its various forms. Following this, the implications of tolling and depreciation methodology in a COS framework are discussed. An evaluation of COS regulation is then presented.

After a general discussion of incentive regulation, various incentive alternatives are described in Section 5. These include: price caps, automatic rate adjustment mechanisms, yardstick competition, sliding scale and profit sharing plans, banded rates of return, benchmarking, and capital cost incentives. Other schemes aimed at improving pricing efficiency, such as priority pricing, optional tariffs, anonymous mechanisms, and franchise bidding, are also outlined.

An assessment of these various incentive schemes is presented in Section 6. Included here are a number of case studies outlining the experience with incentive regulation applied to other industries or in other jurisdictions. Then, the various schemes described in the previous section are assessed using the same regulatory objectives and criteria as employed in the evaluation of COS regulation.

A summary and the main conclusions are presented in Section 7.

2 *Role and Characteristics of Gas Transmission*

The objective in this section is to highlight a number of background issues relevant to the regulation of long-distance gas pipelines in Canada. The first section provides an overview of the development and role of the gas transmission industry. Key characteristics of this industry from a regulatory viewpoint are set out in Section 2.2. Section 2.3 provides a discussion of some of the main changes in the regulatory and market environment.

2.1 THE DEVELOPMENT OF THE NATURAL GAS INDUSTRY AND THE ROLE OF PIPELINES

The origins of Canada's natural gas industry can be traced to the late 1800s. Gas and oil seepages southwest of Calgary were first noted in a geological survey in 1870. Significant quantities were discovered near Medicine Hat in 1883 and 1890 in the process of drilling for water and coal deposits and by 1909 both Medicine Hat and Calgary had street lighting fueled by natural gas. A few years later, in 1912, a pipeline was constructed to transport gas from Bow Island to Calgary, a distance of about 170 miles.[6]

Large quantities of natural gas were also found with the Turner Valley oil discoveries in 1914 but these were mostly flared. Then, following the Leduc No. 1 oil discovery in 1947, there was a series of major oil finds with substantial amounts of associated gas. Large gas reserves were also discovered during this period,[7] but like the gas associated with oil production, they were viewed at the time as having little value. In most cases there were no significant populations close to the gas supplies to provide a market and it was uneconomic to construct the pipelines and

6 A convenient summary of the development of the oil and gas industry can be found in Petroleum Resources Communication Foundation (1992a, 1992c).

7 For example, substantial gas reserves were discovered at Cessford in 1947, at Pincher Creek in 1948 and at Fort St. John, British Columbia in 1951.

other infrastructure necessary to transport it to and distribute it within the larger, more distant population centres. Until the formation of the Turner Valley Gas Conservation Board (which became the Oil and Gas Conservation Board in 1938) most of the gas produced in Alberta was flared.

Over the past four decades, natural gas has gone from being largely an unwanted by-product of oil production to a highly valued resource which meets the growing domestic and export demands for clean and efficient energy. Total Canadian gas production increased more than fifty-fold over this period, from about 72.6 Bcf (or two billion cubic metres) in 1952 to over 4.5 Tcf (or 128 billion cubic metres) in 1993.[8] At present, there are about 700 companies producing natural gas in Canada.[9] These range from large multi-national corporations to small, emerging producers. Fifteen large and integrated companies typically account for just over 50 percent of total production and the fifteen largest intermediate producers account for another 20 percent.

Associated with the growth in natural gas production has been the development of many important industries. Most of these are related to exploration and development activities, gas processing and the further up-grading of natural gas by-products, the transmission of natural gas to export and domestic markets, and the distribution to residential, commercial, and industrial customers. In most of these, large economic impacts occur through the substantial capital and operating expenditures, the high degree of linkage with other industries, and the significant contributions to exports or import replacement.[10]

Pipelines have played a key role in this development of the natural gas industry. In most cases, the product can only be economically transported by pipeline. Prior to the 1950s there were no major pipelines to take Canadian gas supplies to markets beyond the main population centres in close proximity to the producing fields. However, this would change quickly with the completion of a number of pipeline projects, particularly those associated with Alberta Gas Trunk Line Co. (AGTL,

---

8 Data from Canadian Petroleum Association, Statistical Handbook, Section III, Table 10 and National Energy Board (1993b, 15).

9 This and other data in this section are from The Canadian Gas Association (1992), various issues of Oilweek, and Statistics Canada (Cat. 26-201 and 26-202).

10 For example, in 1991 total investment in industries related to natural gas amounted to $22.8 billion. Of this, $2.5 billion was for gas pipelines, $0.8 billion was for gas distribution and $0.7 billion was for gas processing. Other Statistics Canada estimates of the various dimensions of the industry are summarized in Canadian Gas Association (1992).

now NOVA), Westcoast Transmission (Westcoast, now Westcoast Energy Inc.), TransCanada PipeLines Limited (TCPL), and Pacific Gas Transmission Company (PGT).[11] The construction of these large gas pipeline systems provided access to diverse and distant markets and the resulting growth in demand provided the impetus for the rapid expansion of upstream activities.

This process has continued with numerous expansions to the original transmission systems and the completion of new ones. The latter include the Foothills Pipe Lines Ltd. system (the Canadian Prebuild of the Alaska Natural Gas Transportation System) which initiated flows on the western leg in October 1981 and on the eastern leg in September 1982, and the completion of the Trans Quebec and Maritimes Pipeline Inc. system to Quebec city in late 1983.

Recent expansions, especially for the TCPL, Westcoast and NOVA systems, have been particularly significant. Between 1990 and 1992, $1.7 billion worth of additions were made to the NOVA system.[12] The National Energy Board (NEB) alone has approved gas pipeline expansion projects valued at another $3 billion over the same period and many additional projects and expansions are planned for the near future.[13] These main transmission systems and the associated net plant investments (or rate bases) are indicated in Figure 2.1.

2.2. FUNDAMENTAL CHARACTERISTICS OF GAS TRANSMISSION

There are various attributes of the Canadian natural gas transmission industry that are important in any discussion of regulatory regimes. These are summarized below.

Capital Intensity. Natural gas transmission is a very capital intensive activity, whether measured in terms of capital per unit of output or

11 These included the formation of the Alberta Gas Trunk Line Co. in 1954 (now called NOVA) to gather and transport gas within Alberta, the completion of the Westcoast Transmission line in 1957 to carry Alberta and B.C. gas to markets in Vancouver and the U.S. Northwest, the operation of the TransCanada PipeLine system beginning in 1958, which moved Western Canadian gas to markets in Central Canada and the U.S., and the completion, in 1961, of the Pacific Gas Transmission system, which carried Alberta gas to California and U.S. Northwest markets (the Canadian portion of this system now consists of facilities owned and operated by NOVA and Alberta Natural Gas Company Ltd). The first exports were carried by the Canadian-Montana Pipe Line Company in 1951. See Plourde (1986).

12 Data from NOVA Corporation of Alberta, *1992 Annual Report*, (19).

13 Data from National Energy Board (1990c, 1991, and 1992c, Appendix C).

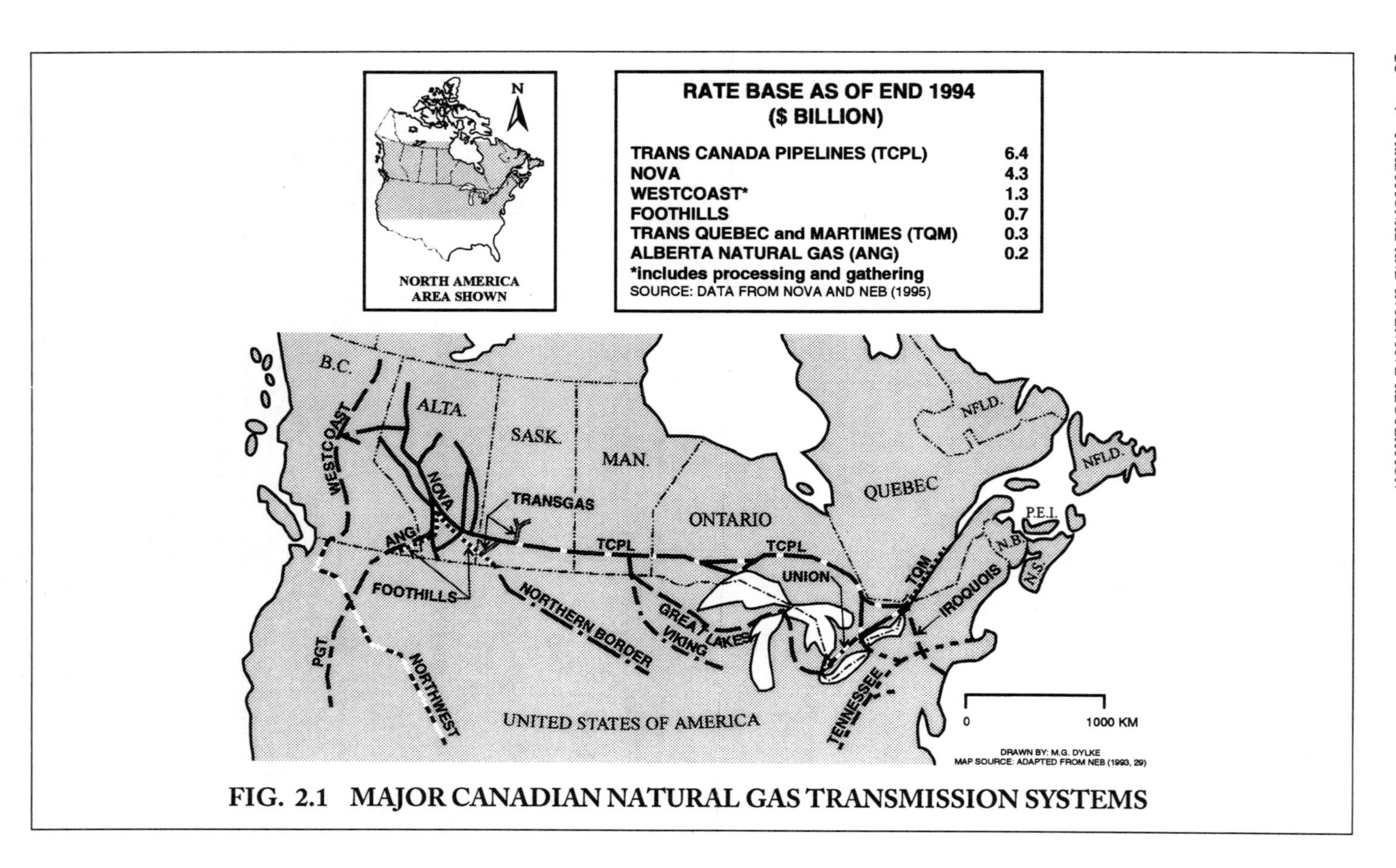

FIG. 2.1 MAJOR CANADIAN NATURAL GAS TRANSMISSION SYSTEMS

capital per unit of labour. One consequence is that, except for highly depreciated systems, the cost of capital is a major component of the cost of gas transportation. Because of the large increments to Canadian pipeline capacity added since the mid-1980s, combined with the fact that the system was relatively young and undepreciated to begin with, the cost of capital portion of transportation costs (including capital recovery or depreciation) represents a large proportion of total costs. For example, the return on rate base plus depreciation in 1993 for the four largest Canadian gas pipelines averaged 54 percent of their total revenue requirements; the percentages for the individual pipelines are shown in Figure 2.2.

It is also useful to note that these proportions of total costs are typically much higher than is the case for most U.S. gas pipelines. The opposite side of this is that operating and maintenance costs for Canadian pipelines are generally a much smaller percentage of total costs than is the case for the more mature and static U.S. pipelines.

These characteristics are likely to remain given the projections of continued significant growth in Canadian gas pipeline capacity, particularly for export markets.

There are a number of implications that should be emphasized. First,  once the investment is in place, the capital cost component is largely a function of exogenous factors such as pipeline age and interest rate levels and, as such, it is mostly beyond the control of pipeline company management. Consequently, these costs are largely insensitive to the productivity incentives implicit in many incentive regulation alternatives. Further, unless the regulatory mechanism allows changes in exogenously determined costs, such as those arising from variations in interest rates, to be passed through to customers, the effect is to expose the pipeline to significant additional risk.

Second, tolls for the major Canadian gas pipelines will be particularly sensitive to the cost of capital. For example, any shifts in regulatory regimes or policies that place these pipelines at greater risk and, therefore, increase the cost of capital, could have a very significant adverse effect on tolls and the competitiveness of Canadian gas in export markets. This increased risk may also mean a move to higher equity ratios for the capital structure which, given that equity is usually more costly than debt, would also raise total costs.[14]

---

14 The much higher equity ratios for U.S. pipelines than for Canadian pipelines is often explained by the fact that pipelines operating under the U.S. regulatory system are exposed to greater risks than is the case for pipelines operating within the Canadian regulatory system.

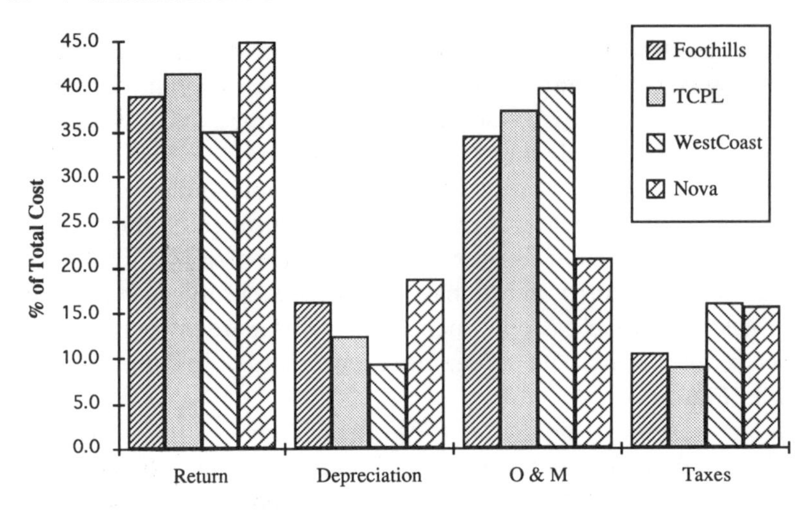

Data from Appendix F2 of NEB (1993b) and NOVA Corporation of Alberta, Annual Report.
Note: Figures for Westcoast include costs associated with gathering and processing.

FIG. 2.2 BREAKDOWN OF COSTS FOR MAJOR GAS PIPELINES

Third, of all the costs, those for operating and maintenance are really the only ones that are directly controllable by the pipeline company in the short run. While these costs might legitimately be the focus under some incentive regulation schemes, they typically account for a smaller proportion of total costs in the case of the major Canadian pipelines than the costs associated with capital (including depreciation). As shown in Figure 2.2, operating and maintenance costs represent about 20 to 40 percent of total costs but typically less than half of these would be subject to managerial discretion. Many elements of these costs would not be variable given reasonable safety and service standards. Since operating and maintenance costs are a larger proportion of total costs for most U.S. pipelines and because the rate of capacity expansion is lower in the U.S., the case for some incentive or non-traditional regulatory schemes may be easier to make in the U.S.

Finally, the importance of capital costs suggests that the focus should be on effective regulatory constraints on the cost and size of expansions. The effects of different regulatory systems on these decisions are likely to be considerably more important factors in overall efficiency than any effects on decisions with respect to the operation of the facilities. Well designed incentive schemes which control the costs and size of capacity expansions may well deliver substantial cost savings.

Durability and the Sunk Nature of Capital. The expected life of pipeline facilities is long, with some components having expected service lives of more than 30 years. This has two important implications. The first is that it would appear to circumscribe the possibility of rapid technological improvement. The second is its effect on incentives for investment.

To the extent that process innovations are embedded in new capital investment, the long-term nature of capacity investments reduces the scope for innovation. Moreover, the technology of gas transmission appears to be much more stable and mature than that in industries such as telecommunications where substantial regulatory changes have occurred in response to rapid technological change. Furthermore, the movement towards deregulation in other industries is due not only to rapid technological change, but to the nature of that change.

Technological changes in telecommunications, especially in long-distance communications, have reduced the significance of economies of scale and allowed for the emergence of a reasonably competitive market. To the degree that technological change in gas transmission has favoured larger diameter pipelines and larger compressors, the effect has been to increase rather than decrease the significance of economies of scale.[15] To the extent that future technological changes in gas transmission will involve the introduction of better information technology that entails large fixed costs, such changes will once again favour large-scale over small-scale operation.[16]

Investments in pipe, and to a lesser extent, in compressors, are sunk expenditures since the opportunity cost of the capital *ex post* is significantly less than its opportunity cost *ex ante*. This means that most of the facilities are specific and cannot be redeployed to some other activity. Assets which can be redeployed into another activity are likely only transferable after a significant write down.

Producers, intermediaries such as Local Distribution Companies (LDCs), and ultimate gas consumers also make large investments which have the characteristics of being sunk and irreversible. This, combined with the same characteristics for the transmission facilities, can lead to a 'hold-up' problem where each party must be protected from the ability

15 Economies of scale exist if average costs decline as the quantity of output increases. The extent of economies of scale in gas transmission is considered in the next section. The relationship between economies of scale and the impetus to regulate is considered in Section 3.

16 A fixed cost is a cost which does not vary with the amount produced and is a source of economies of scale. See the next section.

of the other to act in such a way that the recovery of capital is threatened.

The hold-up problem can be more fully explained using the case of a pipeline and a group of producers. Investments in wells and pipelines are irreversible sunk expenditures which typically lock parties into a long-term relationship. The use of short-term contracts or spot markets to govern the relationship will often not work very well since each side has an incentive to try to 'hold-up' the other side or act opportunistically. This manifests itself when one side has more alternatives or opportunities than the other. The side with more outside opportunities will threaten to leave the relationship, leaving the investment of the other side stranded, unless the other side agrees to new terms. The sunk expenditures of the other side make it difficult for it to resist making concessions since, if it were to exit the relationship, it would forego the possibility of recouping these costs. One way to limit this kind of behaviour is through long-term contracts that legally bind the two parties together and restrict the possibility of exit from the relationship.

The pipeline company knows, prior to the producers having committed investment expenditures to develop a field, that if it raises tolls too high it will limit development. Development requires that the producers be able to anticipate a netback which will cover all costs. However, after the field is developed the pipeline company knows that producers will pay tolls up to the point where their netback equals their variable costs, since in the extreme case they have no other means to get their gas to market and they have to worry about drainage.[17] Of course, producers should and would anticipate the expropriation of their investment and, in the absence of a long-term contract which guarantees access to the pipeline at reasonable prices, are unlikely to invest optimally. They will either decide not to invest in exploration, drilling and development or, more likely, they will underinvest in finding and developing reserves of natural gas.

On the other hand, the profits of the pipeline depend on throughput. This provides producers with the opportunity to hold-up the pipeline. If producers can credibly threaten to make alternative arrangements or to forego development expenditures, they can negotiate with the pipeline for lower tolls and expropriate the sunk capital investment of the pipeline. The pipeline will agree to any toll provided it exceeds the

---

17 Drainage refers to loss of reserves due to a failure to extract the gas at a sufficient rate. For example, given the migratory nature of gas, reduced production by one party may mean some of its reserves will be captured by other companies continuing to extract gas from the same formation.

variable costs of transmission. Anticipating this, pipelines will be reluctant to invest unless they have assurances that producers will commit to use their pipeline at prices and volumes which ensure a reasonable return on investment and a recovery of the invested capital.

The 'hold-up' problem can be, and typically is, *partially* mitigated through the use of long-term contracts which legally bind the parties over a significant portion of the expected life of the facilities. However, long-term contracts do not provide complete protection against the hold-up problem because they are incomplete. In these circumstances, an important role of the regulator will be mitigation of the hold-up problem through mediating the changes or events not covered by the long-term contracts.

It is useful to distinguish between *complete* and *incomplete* contracts. A complete contract specifies the obligation of each party in every possible circumstance and is enforceable by the Courts. For any set of circumstances, a complete contract states the actions required by both sides to the contract and what the financial terms will be. These contracts provide a mechanism whereby each side can commit to a particular behavior for each set of circumstances that might arise. In essence, their behavior is made predictable by a complete contract since it ensures that each side will find it in its interest to behave as promised.

However, a major problem is that complete contracts do not usually exist. Rather, they will usually be incomplete for any of the following reasons.[18]

(i) *Unforeseen Contingencies.* It may not be possible for the parties to the contract to foresee all of the circumstances which might arise and, hence, they cannot include them in the contract.

(ii) *Too Many Contingencies.* Some possibilities can be foreseen, but the probability of such occurrences is felt to be sufficiently small that it is not worth the cost of considering them explicitly in the contract.

(iii) *Precision of Language.* The terms of a contract are often susceptible to differing interpretations. These differing interpretations arise due to the ambiguity of language.

(iv) *Monitoring Costs.* The relevant circumstances may have been specified in the contract, but each side must incur monitoring costs to determine which contingencies are relevant.

---

18 The following is adapted from Church and Ware (1995). For a more general discussion see Milgrom and Roberts (1992) and Williamson (1985).

(v) *Enforceability.* In order to be enforceable by the Courts, the provisions of the contract must be verifiable.

The case of the California Public Utilities Commission ordering Pacific Gas and Electric Company to change its supply arrangements in Alberta provides an illustration of contracts which are incomplete because of an unforeseen contingency. The existing arrangements involved many producer contracts which did not have specified load factors or take commitments even though there may have been certain implicit or informal promises. A second example related to the costs of monitoring and enforcement would be if a group of producers contracted with a pipeline for transmission service where the terms of the contract involved tolls based on the average cost of transmission. If either the pipeline's costs or its throughput were not easily observable and verifiable to a third party like a court, then the pipeline clearly would have an incentive to misrepresent either its costs or throughput in order to raise price and, hence, profits.

Incomplete contracts do not necessarily bind either party to do as they promised since both sides will have incentives to hold the other up after specific investments have been made. When contracts are incomplete, the other side is never sure if the need for adjustment is a legitimate request to adapt efficiently to an exogenous change or an attempt to strategically renegotiate the contract to hold it up: that is, to expropriate rents or some portion of the value of its sunk investment. To the extent that the hold-up problem cannot be completely mitigated by *private* long-term contracts or the regulatory regime, the levels of investment by consumers, LDCs, producers, and pipelines will not be efficient.

Natural Monopoly. Natural gas transmission is often characterized as a natural monopoly. This is a case where costs are minimized by concentrating production in a single firm rather than having two or more firms competing for the market. The technical condition required for natural monopoly is called *subadditivity.* Following Sharkey (1982), we can distinguish between factors which contribute to subadditivity at the level of the plant (a single pipeline) and at the level of the firm (a multi-pipeline firm or a network).

Plant subadditivity refers to cost efficiencies based on the technology of production and it arises from indivisibilities. A production input is indivisible if it is not possible to scale it proportionately with output. Such indivisibilities are responsible for the existence of economies of scale and economies of scope. The extreme manifestation of an indivisibility is a fixed factor. The amount of such an input does not vary as output changes and it gives rise to fixed costs.

Economies of scale refers to the reduction in the unit or average cost as production of a single good or service expands. In the case of a single product firm, a sufficient condition for a firm to be a natural monopoly is that the technology of production be characterized by economies of scale over all levels of output. Since average costs decline as output expands when economies of scale exist, marginal or incremental cost will be less than average cost.

Economies of scope refers to the reduction in costs when a group of products or services is produced by a single firm instead of by multiple producers. This gives rise to multiproduct firms. A production process is characterized by economies of scope if joint production is less costly than producing the products or services individually. In general, *common costs* will give rise to economies of scope. Common costs arise when multiple products or services provided by a firm are produced by the same set of inputs.[19]

The following are examples of indivisibilities which give rise to economies of scale and scope in natural gas transmission.

(i) *Volumetric returns to scale*. As the diameter of a pipeline doubles, its volume goes up by a factor of four, while its surface area increases by a factor of two (Cookenboo 1955). Output is proportional to volume while cost of construction is generally proportional to surface area. Besides giving rise to economies of scale, this factor is also responsible for economies of scope. For example, the construction of one pipeline to provide service from Empress to Winnipeg and another to provide service from Empress to Toronto may be more costly than constructing one pipeline to provide service to both.

(ii) *Construction Costs*. Many installation costs are fixed. That is, they do not depend on the size of the expansion. Consequently, the larger the expansion, the lower the average cost.

(iii) *Right-of-way*. The size of the right-of-way required does not vary proportionately with the size of the pipeline.

Firm-Level Subadditivity arises from the organizational advantages of concentrating production within a single firm. That is, it arises when the transaction costs of organizing production within a single firm are less than those associated with using the market to organize production.

---

19 See Kahn (1988 v.I, 77) for a discussion of common costs.

These network economies which indicate firm-level subadditivity include the following.

(i) *Economies of Fill.* Expansion (installation) costs are minimized if expansions are infrequent and large. An optimal expansion path will trade-off the lower installation costs of large expansions against the cost of excess capacity for some period until the growth in demand catches up to the large increment to supply. An optimal expansion path is more likely to be realized if there is only a single firm. The coordination costs associated with a group of firms providing an optimal expansion path may be considerable.

(ii) *Network Management.* There are advantages to having a network (like a pipeline system) operated by a single firm. These economies arise from the fact that a single operator will be able to manage a system more effectively than multiple firms since optimal network management will involve activities like rerouting, aggregating, and sharing transmission.

(iii) *Network Configuration.* A network constructed by a single firm is more likely to be configured optimally. Optimal management and optimal configuration will minimize capacity costs.

As indicated by Teece (1990), the ability to realize the economies associated with the optimal sequence and size of expansions, network management, and network configuration will often be greatest when the network is controlled by a single firm. The transaction costs associated with achieving these efficiencies using contracts and market exchange can be large.

There have been few attempts to quantify these various economies for specific pipelines or pipeline networks. While tests for subadditivity have been undertaken for some U.S. systems (for example, see Ellig and Giberson (1993)), there have not been any published analyses of this type for Canadian gas transmission systems except for a recent study by Pawluk (1995). Pawluk provides considerable evidence that the TCPL system is subadditive.

Factor Substitutability. Pipelines are characterized by technology which allows only very limited substitutability between factors of production. In the planning stage there is usually some ability to substitute compression for looping or vice versa. This essentially amounts to a trade-off between fuel and capital. However, once the capacity is in place, the substitutability among the main inputs (capital, fuel, and labour) is negligible. This tendency towards a fixed proportions production process greatly reduces the potential for significant increases or decreases in operating efficiency through altering the incentives / disincentives via changes in the regulatory regime.

Growth Rates and Increment Sizes. As discussed in Section 2.3, it is expected that the Canadian gas pipeline sector will remain in an 'expansion mode' for some time to come. This is in contrast to the situation in the U.S. where this sector is more mature and unlikely to see comparable rates of expansion.

Along with the expected frequency of capacity expansions, an important characteristic is that these usually involve large increments. Except in situations where added capacity can be achieved through modest increases in compression, economics dictates fairly large increments with consequent sizable impacts on tolls. In addition, these increments and toll impacts are very 'pipeline specific.' This, combined with the frequency of expansions, means that tolls cannot be set simply through the application of tidy formulas as is often envisaged in incentive regulation regimes.

Control Over Gas Flows. Unlike the situation prior to gas market and price deregulation (where some major pipelines such as TCPL were merchant pipelines), the major pipelines in Canada are all now regulated contract carriers.[20] As such, they no longer have the ability to determine gas flows and rates of utilization. Consequently, incentive schemes which put the pipeline at risk for low rates of utilization will generally have little applicability in a post-deregulation world.[21]

Demand Variability and Service Quality. In the past, demand on Canadian pipelines exhibited large seasonal and cyclical variations. However, in recent years, the rates of utilization have tended to be high and show much less variability.

The reasons for this include the natural tendency for the quantities of gas purchased to show less variability when prices can adjust (as has been the case since the deregulation of gas prices), the development of

20 A regulated contract carrier provides only transportation and on terms set by a regulator. A merchant pipeline provides a bundled transportation and commodity service. See Teece (1990).

21 It might be noted that most of the discussion of incentive ratemaking for pipelines in the United States occurred prior to FERC issuing Order 636. That is, it was based on an industry structure in which pipelines had a significant merchant function. Commissioner Trabandt (1992, 7–10) has expressed doubts about the applicability of many incentive ratemaking schemes in the post-636 world. As will be noted later, such incentive schemes have been implemented for British Gas but it provides transmission for its own gas and also provides local distribution and retail sales.

spot markets and more widespread use of gas storage, and the greater integration of Canada and U.S. gas markets.[22]

These high and stable utilization rates, in combination with the 'bullet line' character of many Canadian pipelines, have meant that the costs of any transmission service interruptions are very high. Consequently, the impact of regulatory changes on 'quality of service,' especially in the form of system reliability, are particularly important in a Canadian context. As discussed later, the traditional regulatory approaches have fostered the achievement of high reliability levels. Some incentive schemes can produce significant tendencies to lower the level of reliability if unchecked.

Carrier Status. One of the issues in the context of regulatory alternatives concerns the carrier status of the pipeline. For example, the role and nature of regulation will depend to some degree on whether the pipeline is a common carrier or a contract carrier.

There are actually three main categories of carriers. Private carriers are merchant carriers. They transport their own gas for sale and provide their customers with bundled commodity and transportation service. Contract carriers provide transmission service for gas owned by others according to a private contract between the pipeline company and the shipper. Government regulation is not an essential characteristic of either private or contract carriers. Regulation of private or contract carriers arises if there are concerns about such things as monopoly pricing and asymmetries in bargaining positions.[23] On the other hand, a common carrier is created by government edict and it, by definition, requires regulation. A common carrier is typically required to provide service to any shipper willing to pay the (regulated) toll. Common carrier status implies that the carrier may not refuse to supply service and that it must serve in a nondiscriminatory fashion. These features imply that when transmission capacity is insufficient, available capacity must be rationed on a pro rata basis across all customers, usually on the basis of their nominated shipping volumes.

The deregulation of natural gas markets in Canada in 1986 was accompanied by the realization that truly effective competition in the market for natural gas required a fundamental restructuring of natural gas transmission. Prior to deregulation, the major Canadian pipeline companies, with the exception of NOVA, provided bundled gas and

22 See De Vany and Walls (1994) for a discussion of these factors.

23 The distinction here is between voluntary and regulated contract carriers. Voluntary contract carriers transport gas on negotiated terms without regulatory interference. Regulated contract carriers transport gas on terms set by regulators.

transmission service to their customers, mostly local distribution companies. In order to establish competition in natural gas, control over both gas sales and gas transmission by the pipeline companies had to be reduced. In 1986, the National Energy Board did this by mandating access to the major ex-Alberta pipelines (NEB 1985, 1986), essentially converting them from private contract carriers to contract carriers. In the United States, FERC Order 636 mandated a similar transformation (FERC 1992).

On the other hand, most major oil pipelines in the United States and in Canada are common carriers, not contract carriers.[24] The reason for the difference in carrier status would seem to relate, at least in part, to differences in the transmission of oil which mitigate the severity of the hold-up problem. Specifically, oil pipelines face considerable competition and they have often been vertically integrated; that is, oil pipelines have been frequently owned by oil producers.[25]

Teece (1986) demonstrates that oil pipelines face considerable competition. There is competition form other modes, especially ocean tankers and barges, competition from other oil pipelines and petroleum product pipelines, competition from other supply basins, and competition from potential new entrants. All of these provide alternative means to supply crude to the market. The large number of alternative sources of supply for crude oil mean that specific customers do not require assured access to transmission capacity through long-term contracts.

In most instances, natural gas consumers have traditionally not had nearly the same number of supply alternatives. Moreover, the costs associated with a failure of gas supply to reach the buyer are typically very high. Consequently, there are significant advantages to being able to reserve transmission capacity over lengthy periods of time and to have a very high probability that all contracted volumes will reach their specific market destinations. In order to mitigate the potential for hold-up, customers will desire long-term contracts which realize these objectives.

Similar considerations also apply to producers. It is economically feasible to truck oil and the relative cost of construction for oil pipelines is considerably less than the costs of natural gas lines, making entry by

---

24 Compare sections 71(1) and 71(2), Part IV of the National Energy Board Act. The Hepburn Amendment to the Interstate Commerce Act in 1906 conferred common carrier status on oil pipelines in the US. See Teece (1986) and Broadman (1987a) for details.

25 In 1976, 95 percent of all crude oil shipments and 78 percent of all refined product shipments in the U.S. were carried in pipelines owned by the 20 largest integrated oil companies (Hansen 1983, 18).

competitors easier. Moreover, many oil pipelines are owned by oil producers. Such vertical integration usually eliminates the danger of hold-up.

In summary, there are a number of characteristics of gas transmission in Canada that are important to consider in the context of changes in the regulatory regime. These include the fact that the main costs are those associated with the financing and recovery of the large amounts of capital involved in gas transmission; investment decisions concerning the timing and size of capacity are a major determinant of tolls and usually much more important than operating decisions once the capacity is in place; toll patterns over time are more a function of exogenous factors such as depreciation, capacity expansions, and interest rates than they are a function of incentives or disincentives for technological progress and innovation; increments to Canadian capacity are frequent and large, and the resulting toll adjustments cannot usually be made simply through the application of a formula; there is limited ability by pipeline management to affect the size and directions of gas flows; and the issues of reliability and quality of service are very important.

2.3 REGULATORY AND MARKET ENVIRONMENT CHANGES

There have been dramatic changes in the regulation of the natural gas industry over the past decade, primarily related to the deregulation of gas prices and markets. In addition, there have been major shifts in general market conditions. An outline of these changes and a summary of the implications for the regulation of gas transmission follow.

The Period Prior to Deregulation. To provide context for the changes in more recent periods, it is useful to note the environment during the preceding era. Over the 1970s and up to the mid-1980s, the natural gas environment was generally characterized by high and rising gas prices, tight regulation of markets and prices and, especially towards the end of this period, significant growth in supply capability.[26]

Excess Capacity. Since gas prices were controlled, most of the market adjustments were in the form of variations in gas quantities or volumes. This, combined with restrictions on exports, misguided policies, and industry over-optimism, resulted in situations of excess transportation (and production) capacity. In some years prior to the deregulation of gas markets and pricing, load factors fell to very low levels on systems which were heavily reliant on export volumes.

---

26 A good survey of conditions and the regulatory environment over this period can be found in MacGregor and Plourde (1987).

Overbuilding Concerns. One important result of this experience is that many shippers remain concerned about 'over-building' even though the excess capacity in this earlier period was due to artificial or non-market factors, such as regulated pricing and other regulatory controls. This experience also provides the lesson that large fluctuations in volumes are only likely when prices are controlled. If gas prices are allowed to reflect market factors, volumes will exhibit greater stability since more of the adjustment to demand or supply shifts will be via changes in prices and less will be via changes in volumes.

Responsiveness of Gas Supply. Perhaps the most important implication of this earlier experience has to do with the responsiveness of gas supply to market signals. At the time, the standard view of governments and regulatory agencies was that Canada's natural gas reserves were fixed and more or less known. Consequently, any increase in production was seen as moving closer to the present time when there would be falling gas production levels and gas shortages.

The rapid increases in reserves and deliverability in response to the rise in prices served to discredit this view of natural gas as a fixed and known resource.[27] As will be noted later, even with only modest increases in gas prices, gas production is expected to continue to increase at very significant rates (approximately 3.5 percent annually) for the foreseeable future.

Importance of Transportation Costs. During the boom years there were good prospects for higher prices and returns for producers. Higher transmission costs could be more easily absorbed in this environment, especially given the associated expansion in gas supply, and there was less concern about gas transmission efficiency and costs. The cost of transmission over this period was a much smaller proportion of the delivered cost of gas than would be the case in later periods.

Market and Price Deregulation. The process of deregulating natural gas markets and prices began in Canada with the signing of the 'Halloween' agreement in October of 1985.[28] This provided for prices and market allocations of natural gas to be primarily determined by the interaction of gas buyers and sellers in active gas markets that were

---

27 For example, in 1975 the NEB's estimate of Ultimate Potential Gas Reserves was about 119 Tcf (125 EJ). This has steadily increased over time; the NEB's current estimate is approximately 242 Tcf (255 EJ).

28 This agreement, formally referred to as "Agreement Among the Governments of Canada, Alberta, British Columbia, and Saskatchewan on Natural Gas Markets and Pricing" (October 31, 1985), followed the "Western Accord" signed on March 28, 1985.

expected to develop. Part of this process involved the conversion of the merchant pipelines to contract carriers and the opening-up of access to these lines. Also, there were significant changes to some components of the regulatory system. For example, rigid tests and rules regarding gas prices and market allocations were generally replaced with mechanisms that were more sensitive to market conditions. It should be emphasized that there was no intent to deregulate the transmission component of the natural gas sector. Rather, it was recognized that the fundamental reasons for regulating it were still valid. The main objective was to deregulate those components of the natural gas sector for which (as explained by Pierce (1990)) there were no sound fundamental economic reasons for regulation in the first place.

There have been a number of other changes associated with gas market and price deregulation that can be briefly noted. These include modifications to gas supply contracts to incorporate market-sensitive pricing, to provide for greater flexibility and supply basin diversity, and to cover a broader range of contingencies (such as those related to the inability of purchasers to take delivery of the contracted volumes or the inability of producers to meet supply commitments); the general move to shorter term contracts and the development of sizable spot markets for natural gas; expansions in the roles of LDCs and other direct purchasers to include responsibility for transportation arrangements on systems connected to the supply basin and the negotiation of gas supply contracts with supply aggregators or with a variety of individual producers; changes in the roles for many gas producers to include gas marketing and contracting for transportation; the emergence of hubs, the development of markets for gas futures, and the expansion of arbitrage; and, the reductions in the concentration of ownership of transmission capacity rights as well as the development of active markets for secondary capacity.

Gas market and price deregulation in the U.S. has followed a similar pattern to that in Canada but has proceeded at a slower pace.[29] Only with FERC Order 636 has the U.S. reached the final stages of this process, stages that were reached some time earlier in Canada (FERC, 1992).

Declining Energy Prices. As shown in Figure 2.3, real natural gas prices began a significant decline after the mid-1980s. This was the result of a combination of factors, including the dramatic drop in oil prices in late 1985 and early 1986, the excess gas supply situation (the

29 See Pierce (1990) for a detailed history of natural gas regulation and deregulation in the U.S. Also, see Teece (1990).

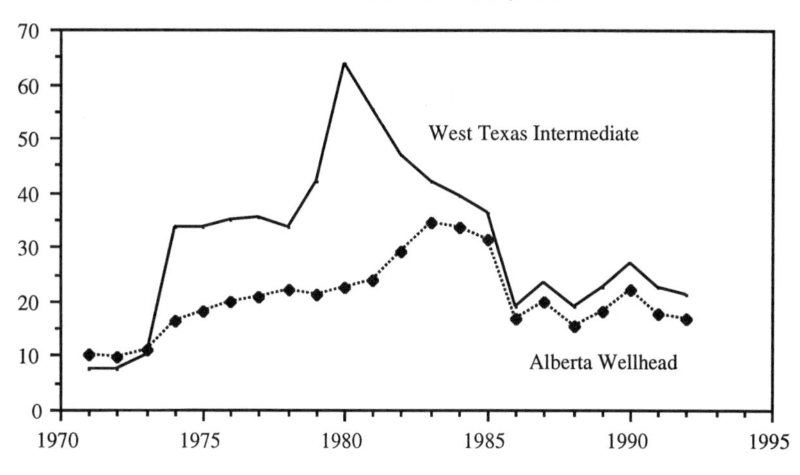

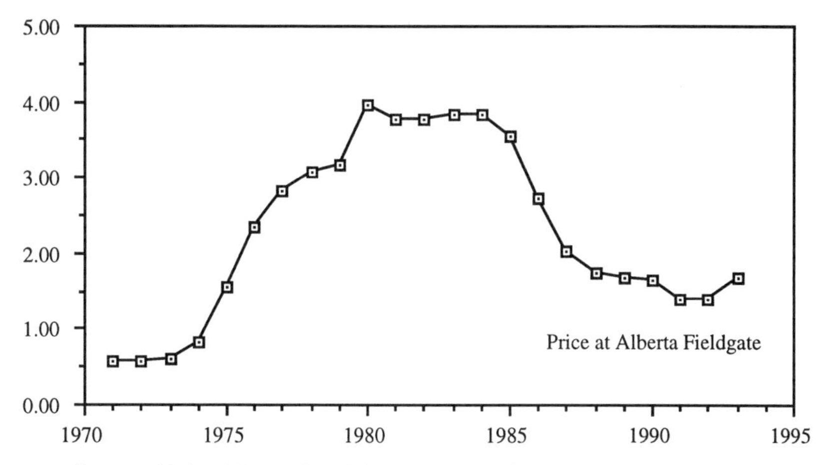

Sources : National Energy Board Canadian Energy Supply and Demand 1990-2010 and Canadian Association of Petroleum Producers Statistical Handbook

FIG. 2.3 OIL AND GAS PRICES

gas bubble) related to prior policies and expectations, and the removal of price controls. The current level of real gas prices is about what it wasin the mid-1970s and approximately one-third of the level it was in the late 1970s and up to 1985.

Increases in Exports. One important effect of the drop in prices was that Canadian gas supplies once again became competitive in export markets and volumes moving to these markets quickly recovered.

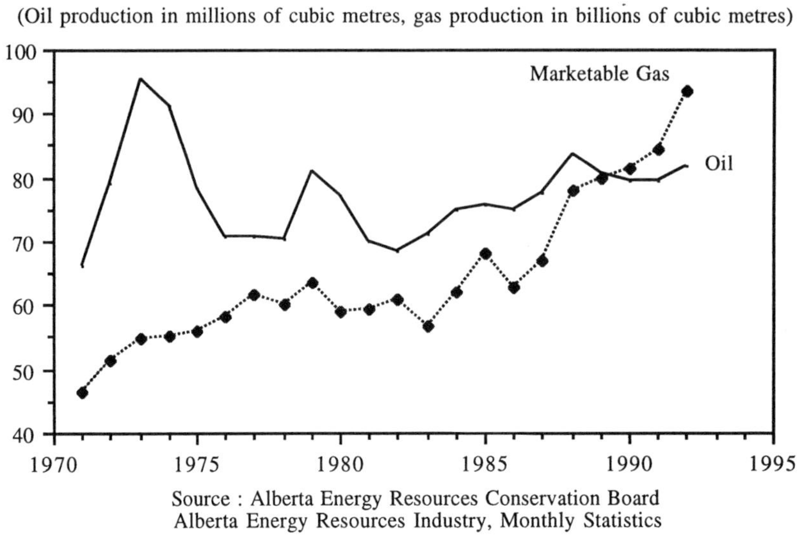

(Oil production in millions of cubic metres, gas production in billions of cubic metres)

Marketable Gas

Oil

100
90
80
70
60
50
40

1970 1975 1980 1985 1990 1995

Source : Alberta Energy Resources Conservation Board
Alberta Energy Resources Industry, Monthly Statistics

FIG. 2.4 PRODUCTION OF OIL AND NATURAL GAS —ALBERTA: 1971–1992

Further, this drop in prices, combined with other factors,[30] paved the way for rather dramatic growth in Canadian gas exports and gas production over most of the period since the mid-1980s.[31] Since 1986, total deliveries of Canadian natural gas have increased by about 10 percent annually (see Figure 2.4 for Alberta production) and exports have increased by about 29 percent annually.

It is reasonable to expect Canadian gas production and deliveries to continue to grow at a significant rate. For example, the NEB's 1994 assessment suggests that by the year 2006, assuming current technology, annual marketable gas production will increase to over 6 Tcf (from about 4.5 Tcf in 1993) and annual exports will increase to about 3.4 Tcf (from about 2.2 Tcf in 1993).[32] Nevertheless, these rates of increase will

30 These include the declines in the value of the Canadian dollar relative to the U.S. dollar, the large supply overhang in Canada, the substantial reductions in costs achieved by the Canadian natural gas industry after the collapse in energy prices, and the greater level of maturity (and higher supply costs) for most U.S. supply basins.

31 For example, deliveries of Canadian gas increased from 2.7 Tcf (75 billion m^3) in 1986 to 4.5 Tcf (127 billion m^3) in 1993, an increase of almost 70 percent over seven years. Gas exports went from 1.1 Tcf (21 billion m3) in 1986 to 2.2 Tcf (63 billion m^3) in 1993, an increase of 200 percent. Data from National Energy Board (1986, 1993b).

32 These figures are for the main case with current technology. See National Energy Board (1994b, Ch. 6).

be much lower than those experienced in the recent past. The lower rates reflect the fact that many of the conditions which allowed such large increases in production since the mid-1980s are unlikely to exist in the foreseeable future.[33] One implication is that, while the demand for transmission capacity will continue to grow, it is unreasonable to expect that the rate of expansion will be anything near that experienced over the period since the mid-1980s.

2.4 DEMANDS FOR REGULATORY CHANGE

There has been considerable pressure in recent years for regulatory changes related to the gas transmission industry. Most of these demands have been by shippers who feel that pipelines are inefficient and could lower transportation costs if the proper discipline or incentives were provided. More generally, there is a spillover of the 'regulatory revolution' that has been underway with respect to other utilities such as telecommunications and electricity.

Producer Cost Reductions. A common complaint by producers is that they have been forced by lower gas prices and increasing competition for markets to greatly improve efficiency and reduce costs but there have been no equivalent changes in transportation efficiency and costs. While it is certainly true that the producing sector has greatly reduced its costs from the levels observed a decade or so back, much of this can be related to rent capitalization. That is, there were sizable economic rents associated with oil and gas production during the period of very high energy prices.[34] Some of these rents became capitalized in the prices for inputs such as land and drilling rigs. When energy prices fell and the rents diminished or disappeared it would be expected that the prices for many inputs would also fall. Further, it is generally true that boom-type conditions lead to decreasing efficiency which is only reversed in the subsequent bust.

This points out an asymmetry between changes in producer efficiency and costs on the one hand, and pipeline efficiency and costs on the other hand. Since rates of return and tolls were tightly regulated

---

33 These conditions include the existence of a large supply overhang, considerable excess capacity (including transmission capacity) and high real prices such that declines can significantly increase demand.

34 In simple terms, economic rent is that amount earned which is in excess of the minimum required to keep the resources employed in a particular activity. For example, if a particular sector typically must earn 15 percent to provide a competitive return to the owners of the capital, and if that sector actually earns 25 percent, the difference of 10 percent would be economic rent.

during the period of high and rising energy prices, the potential for large transmission rents and rent capitalization in the input costs of pipelines was limited. Consequently, when energy prices fell there would not be the possibility of dramatic reductions in input costs.

Also, in the case of pipelines there was no proliferation of marginal organizations which could only survive in an artificial or inflated environment as was the case for the producing sector. In general, the argument that pipelines must be able to greatly reduce costs and tolls because the producing sector has been able to greatly reduce its costs is not convincing. However, this should not be interpreted as an absence of potential for efficiency gains and cost reductions in the pipeline sector.

Regulatory Change for Other Utilities. Another common argument is that dramatic changes have taken place in the regulation of some other public utilities and that similar changes may be appropriate for gas pipelines. Perhaps the best example of major change has been with regard to the regulation of telecommunications, where there has been a pronounced move toward deregulation. However, most of this has been the result of technological changes which destroyed the fundamental reasons for regulation. Specifically, in many areas the new telecommunications technology does not involve the network economies and the economies of scale and scope which formed the foundation for the 'natural monopoly' in earlier periods that was the rationale for tight regulation. These changes have made competition a viable alternative.

Many of the new regulatory regimes such as price caps that have been used for this sector are often advanced as being equally applicable to the Canadian gas transmission system (for example, see the PRIDE proposal (Imperial Oil 1994)). However, it should be noted that these regimes were typically put in place in the case of telecommunications as a transition mechanism until sufficient competition emerged to be able to eliminate most regulation. They were not intended as permanent fixtures.

There have also been major technological and other shifts in the generation of electricity. Whereas large, centralized generation facilities were the most economic way in the past to produce electricity, today the use of small, dispersed facilities (such as gas-fired cogeneration plants) is more efficient in many cases. Again, the requirements for the regulation of generation (although not for transmission) have been modified by technological change.[35] An important point is that there

35 For example, see Alberta Energy (1994).

have been no such dramatic changes in the technology of gas transmission and none are on the horizon. Further, as noted earlier, in many instances the technological changes applicable to gas transmission have served to increase economies of scale and scope rather than, as in the case of telecommunications and electricity generation, reduce these economies. In general, then, the arguments that there should be a parallel between regulatory changes with respect to other utilities and regulatory changes with respect to gas transmission are oversimplifications. In fact, recognizing the correspondence between electricity transmission and gas transmission, along with the acceptance that electricity transmission (as opposed to electricity generation) continues to require regulation, suggests that if there are parallels, they constitute arguments *against*, rather than for, substantial changes in the regulation of gas transmission.

Nevertheless, it is useful to note that the regulatory changes with respect to electricity can have important implications for market competition between gas and electricity. To the extent that electricity prices fall as a result of regulatory changes in that sector, there may be increased downward pressure on gas prices and costs in those markets where gas and electricity are close substitutes.

Other arguments are based on changes in the regulation of pipelines in other jurisdictions. For example, there has been a shift to a form of 'price cap' regulation for some oil pipelines in the U.S. and there is expected to be a general move to greater reliance on competitive pressures to control rates on both oil and gas pipelines in that country. However, it is not clear that these types of changes can be directly applied to major Canadian gas pipelines.

First, unlike the case in the U.S. where there is a significant degree of 'pipeline on pipeline competition' in many markets,[36] this is quite rare in Canada. Second, it is worth noting that 'light-handed' regulatory schemes, such as those used by the NEB for Group 2 pipelines, are already in place in Canada. Many aspects of these schemes are similar to the incentive schemes (such as price cap regulation for oil pipelines) recently proposed in the U.S. (under FERC Order 561). Third, as noted earlier, there are substantial differences between major Canadian gas pipelines and their U.S. counterparts in terms of cost structure (particularly with respect to the relative size of the capital cost and capital recovery components) and in the expected future rate of expansion. For instance, it may be reasonable to use an index pricing system for setting

---

36 See Gallick (1993).

tolls on a mature system where capital recovery costs are minimal and where operating and maintenance costs are the major costs. It may be quite a different matter for a less mature system such as a major Canadian pipeline where costs and tolls are much more dependent on rates of expansion and capital costs (including capital recovery) than they are on the more predictable operating and maintenance expenses. In addition, it is expected that there will continue to be significant expansions of the Canadian gas transmission system. Many of the schemes applied elsewhere tend to be less appropriate in cases where the system is subject to frequent and sizable expansions.

Other Complaints. Various other arguments for significant changes in the regulatory system have been advanced by some shipper groups. As an example, the following are highlighted in a recent proposal referred to as PRIDE.[37]

(i) The environment is much more competitive. A regulated world develops a culture that is alien to the requirements of competition; traditional regulation tends to rigidity and conservatism; and, most sectors must become more competitive and efficient to survive.

(ii) Regulatory changes in the U.S. will increase competitive pressures on Canadian oil producers.

(iii) The pipeline industry has matured. No further major expansions will be necessary.

(iv) There is frustration with traditional regulation. The process of scrutinizing every pipeline management decision is tedious and it is often ineffectual and frustrating.

(v) There is a growing body of experience from other regulated industries (such as telecommunications and electricity in the U.K.) that competition works.

(vi) Current procedures for proposing and authorizing expansions result in sub-optimal expansions. In general, the fear of over-building means that expansions are usually too small and delayed too long.

Some of these arguments have already been dealt with. For example, there is a major difference between potential competition in telecom-

---

37 See Imperial Oil Limited (1994).

munications or electricity generation on the one hand and, on the other, potential competition in the case of gas transmission in Canada. Some of the arguments are impossible to address. For instance, it is often suggested that there are major opportunities for efficiency gains and cost reductions in the case of gas pipelines, but no evidence of inefficiencies is presented and no examples of unexploited efficiency gains are given. Arguments that the Canadian pipeline system is not likely to expand much more would not seem to be valid for the major gas transmission systems in Canada.

Some of these arguments reveal inherent contradictions. For example, most shippers argue that there is a tendency for pipelines to overbuild and that this tendency must be tightly controlled. However, at the same time it is suggested, as in the PRIDE proposal, that the under-expansion of capacity is costing the petroleum industry large amounts of money and that the expansions should be more responsive to demands by shippers. Similarly, the recommendations advanced by shipper groups often focus on mechanisms that would put the pipeline shareholders at greater risk concerning recovery of their investments and operating costs. However, this may result in outcomes which are opposite those that they are arguing for. If the investments associated with expansions are at greater risk, the result will likely be fewer and smaller expansions, rather than more and larger expansions. Also, if the pipeline is exposed to greater risk, the increase in the cost of capital may mean higher tolls, not lower tolls.

In summary, there are significant demands for regulatory reform in the case of natural gas transmission. While the arguments for particular types of reforms are often not well-supported, they do, nevertheless, demonstrate a strong view that regulatory reform must be considered. However, before any major reforms are initiated it is appropriate to undertake a careful analysis of the alternatives and tradeoffs. Such an evaluation is the focus in later sections.

3 Regulatory Concepts and Objectives

This section provides a discussion of regulatory concepts and objectives that are relevant to the issue of alternative regulatory regimes. A brief introduction to the main theories of regulation is set out in Section 3.1. The rationale for regulation is examined in Section 3.2. The next section deals with optimal pricing. Section 3.4 presents the standard regulatory criteria for evaluating regimes.

3.1. THEORIES OF REGULATION

A number of general theories are used to explain why certain sectors are regulated.[38] One group falls under the heading of *Economic Theories of Regulation.* These generally take the view that there is a demand for regulation arising from those groups that would benefit from the redistribution of wealth resulting from regulation, and a supply of regulation arising from governments and the political process. This framework is then used to attempt to explain various types of regulation as an outcome of the interaction of supply and demand. For example, it is suggested that regulation will usually be such that it benefits small, concentrated groups more than it benefits large, diffuse (and unorganized) groups.

Another theory set, which is more applicable to the case of pipeline regulation, is that termed *Public Interest Theory.* Here, the rationale for regulation is that conditions of market failure, such as those giving rise to natural monopoly, make an unregulated or competitive outcome unworkable or unacceptable. In this framework, the task of the regulator is to impose entry and pricing constraints so that the outcome is as close as possible to that consistent with the maximization of societal welfare. A further variation falls under the heading *Capture Theory.* According to this approach, regulation is initially implemented based on clear evidence of market failure. However, after the early phase when

38 See Schmalensee (1979) for a survey.

the regulatory decisions serve the public interest, the regulators become 'captured' by those they regulate. Consequently, it is argued that eventually the regulation primarily serves the interests of those who are regulated.

The approach in this study most closely follows that associated with Public Interest Theory. That is, it is assumed that the primary function of regulation in the case of natural gas pipelines is to correct for market failure. For example, for many of the reasons discussed earlier in Section 2.2, it can be argued that an unregulated pipeline industry would not produce acceptable outcomes in terms of such things as efficiency, proper price signals, and equity or fairness. The main issues then concern the relative merits of alternative regulatory regimes in terms of producing outcomes consistent with the public interest.

3.2. THE RATIONALE FOR REGULATION

Regulation is broadly defined to be intervention by administrative agencies through the imposition of rules and actions which either directly affect market outcomes (prices, production, variety, quality, etc.) by changing market institutions, or indirectly affect market outcomes by imposing constraints on the behaviour of market participants (the firms and their customers).[39] The imposition of constraints on market participants changes their incentives and, consequently, alters their behaviour and the market outcomes. The primary intent is to achieve an efficient outcome as well as one which meets equity or fairness criteria.

It is important to distinguish among the various elements of regulation. A common minimum would involve regulation to meet basic criteria concerning such things as the safety of the population and the protection of the environment. Although there may be important issues regarding the appropriateness or effectiveness of certain of these types of regulations, for our purposes they are taken as a given. Rather, the focus in this study is on the primary economic elements involving such aspects as the degree and nature of price regulation and the regulation of capacity allocation and expansions.[40]

Efficiency. It is well known that competitive markets, under certain conditions, result in outcomes which are efficient. An efficient outcome maximizes the sum of producer and consumer benefits from production

39 This definition of regulation follows Spulber (1989, 37).

40 See Berg and Tschirhart (1988) for a good overview of the rationale for economic regulation.

and consumption. An outcome where it is not possible to increase the welfare of one individual without decreasing that of another is termed *Pareto Efficient.*

A dollar measure of the benefits to consumers of having the opportunity to purchase output at a given price is called *consumer surplus*. Consumer surplus is the sum of the differences between the maximum that consumers would be willing to pay for a good or service and the market price they actually pay. The dollar measure of the benefit to producers from production is simply profit: it is the sum of the difference between the market price producers receive and the price per unit that would be just sufficient to induce them to provide the commodity. The minimum they have to be compensated to supply an additional unit is the *incremental cost* or *marginal cost*.

The sum of consumer surplus and producer surplus is *total surplus*. The total surplus on a per unit basis is the difference between the maximum a consumer is willing to pay and the incremental cost. The condition for maximizing total surplus requires production to be carried to the point where the value consumers put on an increment of supply (as reflected by the market price they pay) is equal to the cost of the increment of supply (or the marginal cost). At such a level of production, there are no gains left from voluntary trade. No market participant can find another individual with whom a trade could be made that would make both better off. Consumers would only buy another unit if the price were to fall; producers would only be willing to supply another unit if the price were to rise. An efficient outcome maximizes total surplus, the dollar measure of gains from trade.

The efficiency of competitive markets is due to three features. The first is that, in competitive markets, the level of output is efficient. The value to consumers of the last unit consumed equals the price. The cost to producers of the last unit produced equals marginal cost. Thus, for the last unit produced, the marginal willingness to pay equals marginal cost and, as a consequence, total surplus is maximized. The second feature is that the output produced is rationed efficiently among consumers. Only those consumers whose willingness to pay is greater than the price consume the output. The third feature is that aggregate production costs are minimized. Each firm minimizes the cost of producing its output and the number of firms in the industry adjusts so that each unit of output is produced at minimum average cost.

Components of Efficiency. As the preceding paragraph indicates, there are a number of requirements for an efficient outcome or allocation. These are discussed in more detail below.

(i) *Allocative Efficiency*. This means that the quantity of product or service supplied is efficient. If this efficiency is not achieved, units of output are either not produced, even though consumers are willing to pay more than the cost of production, or too much output is produced, in which case, for some range of the output level, consumers are not willing to pay the full (social) cost of production.[41] For instance, suppose that a pipeline has excess capacity and the incremental or marginal cost of transporting another mcf of gas to market A is $.50 per mcf. Also, suppose that there is currently a profit of $0.75 per mcf, excluding transportation costs, from selling gas in market A. Assuming that the city gate price does not change, a producer would be willing to pay up to $0.75 to ship another mcf of gas to this market. Note that the pipeline and the producer could be made better off by shipping an additional unit. As long as the toll on the extra unit was less than $0.75 but more than $0.50, both parties would gain. These gains would exist as long as the value of transportation to shippers exceeds the marginal cost of transportation. When all such gains from trade are exhausted, the quantity sold is said to be *efficient*.

A failure to achieve allocative efficiency is typically associated with market power. This power arises when competition within the market is weak, usually due to an insufficient number of competitors. The extreme case is when there is only one seller (monopoly) or one buyer (monopsony). When a monopolist raises price above marginal cost, the willingness to pay of consumers also exceeds marginal cost. This reduces total surplus but the monopolist is better off because its share of total surplus rises.

(ii) *Rationing Efficiency.* This concerns the distribution of a fixed level of output (capacity) among consumers. Rationing efficiency occurs when the level of output (capacity) is allocated among consumers such that the value to consumers is maximized. As an example, suppose that 100 mcf / day of transmission capacity is

41 The social cost of production is the cost to society of all the resources used in producing a commodity. This may differ from the private costs of the producing firm if the firm uses resources in the production process for which it does not pay. The most obvious example of a situation in which private costs and social costs differ is when production involves pollution or other environmental costs which are not compensated.

currently being used by a shipper who nets, after transportation costs, $0.50 per mcf. If there exists a second shipper without capacity who could net $0.60 per mcf in the same market, then the capacity has not been allocated efficiently among shippers (the 'consumers' of pipeline services).

(iii) *Cost Efficiency*. This requires that production of a given level output use the minimum amount of resources possible and that the factors of production are used in the proper proportions. If either of these conditions does not hold, the output will not have been produced at minimum cost and, as such, there will be cost inefficiency. One element of this cost efficiency is *managerial efficiency*. This requires the absence of significant managerial slack or organizational inefficiency.

It is the perceived cost inefficiency and the potential for cost inefficiency which tends to dominate the arguments for regulatory reform. For example, it is typically alleged that Cost of Service (COS) regulation creates incentives for excessive inputs (the 'gold plating' argument) and does not create incentives for management to bargain sufficiently 'hard' with input suppliers in establishing input prices.

(iv) *Efficient Product Selection*. This type of efficiency is important if products are differentiated. Pipeline transportation service is differentiated on the basis of reliability, destination, and season. As an example, transportation from Empress to Winnipeg is different than transportation from Empress to Toronto. Similarly, firm and interruptible service are clearly different types of service. This type of efficiency requires that the types of service available match the types of service demanded. For instance, it may not be achieved if, as is often the case, all firm service shippers are provided with the same level of reliability of service. There may exist shippers who would be better off with either a higher or lower level of reliability than that provided to all.

(v) *Cost-Reducing Investments*. This type of efficiency is associated with innovation and investments in cost reductions that are socially efficient in the sense that they are carried to the point where marginal benefits equal marginal costs.

***Market Failure Test*.** The common economic approach to determining the desirability of regulatory intervention is the *market failure test*. Market failure occurs when there are certain characteristics or conditions that prevent the operation of free markets from achieving the efficient and socially optimal outcomes noted above. The first step in the market failure test involves determining that the performance of the particular market when it is unregulated generates socially unacceptable results. Such unacceptable outcomes usually arise because of inefficiencies and / or inequities. As noted later, other steps involve a determination of whether the outcome under regulation (which will almost always be imperfect in practice) would be substantially better than the outcome with imperfect markets (market failure).

In the context of natural gas transmission, there are two main sources of market failure which typically justify the regulation of price and entry. These are: (i) subadditivity and natural monopoly; and, (ii) large sunk specific investments. The existence of both in the case of natural gas transmission was detailed earlier in Section 2.2

(i) *Subadditivity and Natural Monopoly*. If the technology of production is subadditive, the costs of production will be minimized if there is only one firm in the industry. Dividing industry output among two or more firms will lead to higher industry costs than if the same level of output were produced by a single firm. In order to attain cost efficiency at the level of the industry, the number of suppliers should be one. In regulated industries this efficient market structure is created by imposing entry controls: the regulator restricts entry and either explicitly or implicity grants a single firm a monopoly franchise.

Of course, granting a firm monopoly status requires the regulation of prices to control the market power of the franchise monopolist. In the absence of such regulation, the firm will have the ability to set prices (or tolls) above marginal or incremental cost. In order to get a higher price and earn larger profits, the monopolist would restrict capacity. With entry restricted by regulation, the monopolist would not need to worry about alternative sources of capacity competing for the market. The result would be smaller pipelines, less gas transported, and higher tolls than those using an efficient benchmark. One objective of regulation is to prevent outcomes which deviate sharply from the outcomes for this benchmark.

There are two aspects of monopoly pricing which are problematic from a societal perspective and which are often the focus of regulation. The first is the allocative inefficiency associated with a supply or capacity that is less than the efficient supply or capacity.

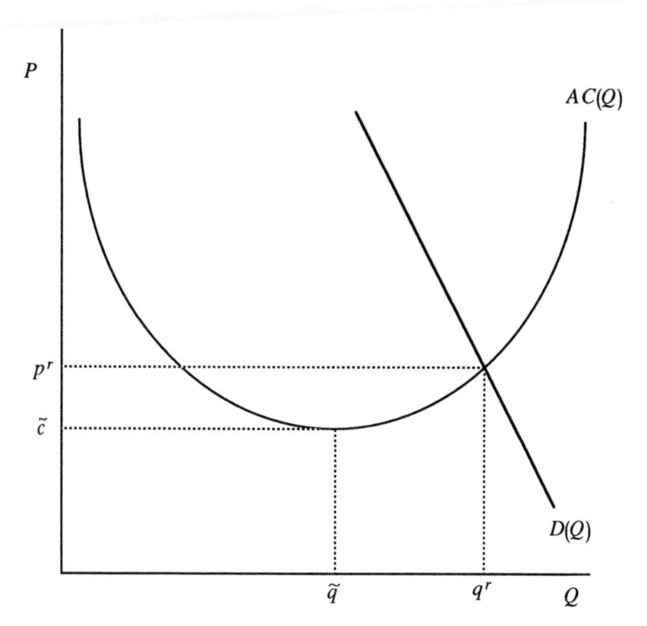

FIG. 3.1 INEFFICIENT ENTRY

At the level of production selected by the unregulated monopolist, the value to shippers of an additional mcf of gas transported (as measured by the price or toll) is greater than the cost of providing one more mcf of capacity. As a result, there is lost surplus and not all the gains from trade are realized. The second problem is that monopoly pricing also involves a redistribution of surplus. The monopoly firm will earn higher profits and shippers will realize less profit on the sale of gas than if transport services were provided efficiently.

To summarize, if the technology of production is subadditive, efficiency is attained by restricting entry and regulating prices to control the resulting monopoly power. An important conclusion is that when an industry is a natural monopoly, *entry* and *competition* are socially undesirable.

Entry is necessarily inefficient since it will raise industry costs, even though it may be privately profitable. Figure 3.1 provides an example of this for a single product firm.[42] The demand curve and long-run average cost function are denoted $D(Q)$ and $AC(Q)$. At the price and output combination p^r and q^r, the monopolist sets prices at average cost, demand is satisfied and economic profits are

42 This example is from Panzar and Willig (1977).

zero.[43] However, this outcome is not sustainable since there is an incentive for inefficient entry. An entrant could enter, charge a price between $\tilde{c}$ and p^r, produce $\tilde{q}$ units of output and make positive economic profits. Notice that in order to make positive profits, the entrant does not produce all that is demanded at the lower price which it charges. By partitioning the market, the entrant is able to realize lower costs and profitably serve some segment of it. Those consumers not served will have to be supplied by the incumbent at a substantially higher cost and the combined losses of these consumers and the incumbent are greater than the profits of the entrant and the increase in surplus of the consumers fortunate enough to be supplied by the entrant.[44]

(ii) *Sunk Specific Investments.*[45] In Section 2.2 it was observed that the high cost of writing and enforcing private long-term contracts to protect capital investments under conditions of uncertainty could negatively affect the incentives of firms to make irreversible investments in facilities. It was noted that after sunk investments are made, each party to a transaction has an incentive to try to renegotiate the terms of the contract and expropriate the capital investment of the other party. Producers could negotiate for lower tolls by threatening to use an alternative (existing or potential) pipeline and the pipeline could refuse to provide service unless tolls increased.

Goldberg (1976) was the first to observe that regulation is, in effect, an 'administered contract' between consumers and the firm. Regulation provides an institutional framework within which adjustment of the terms of a long-term contract between a firm and its customers can occur. Regulation can be interpreted as a set of rules for negotiation and dispute resolution which reduces the transaction costs associated with long-term contracts and sunk capital investments. In particular, regulation provides a means to protect against the hold-up problem by: (i) granting exclusive

---

43 As shown in Section 3.3, this output and price combination is the second-best optimal allocation and p^r is the Ramsey price. It is also useful to note that the concept of economic profits is used here and in the other examples in this section. It will be recalled that if a firm earns a normal level of accounting profit it is said to be earning zero economic profits. In other words, positive economic profits translates into above-average accounting profits and negative economic profits translates into below-average accounting profits.

44 An example of inefficient by-pass when the incumbent firm produces multiple products is provided in Section 3.3 below.

45 The discussion in this section follows that of Spulber (1989, 43–46 and 56–65).

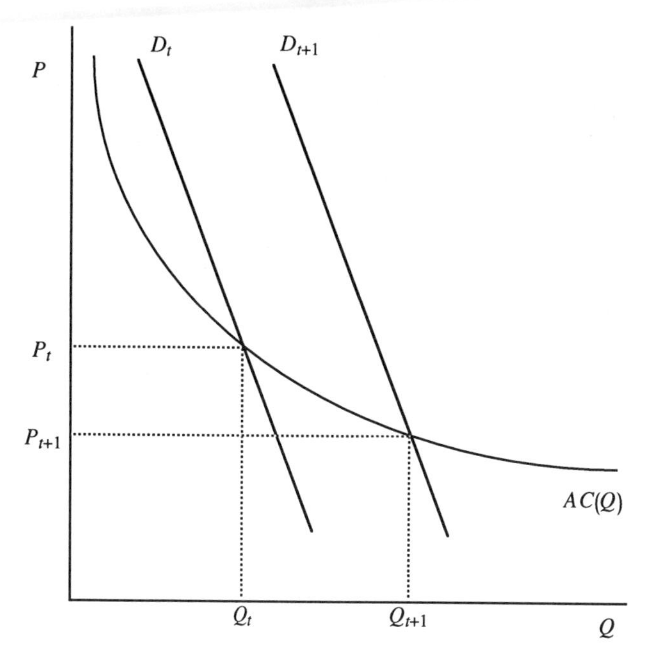

FIG. 3.2 HOLD UP IN A GROWING MARKET

rights to serve (perhaps implicitly) and the rights to customers to be served; and (ii) by providing a cost effective means to adapt the prices, service terms and other contract terms to changing circumstances.

For example, as part of the administered contract the firm agrees that, in exchange for an exclusive right-to-serve and the responsibility to serve the market, it will be granted sufficient revenues to cover its costs, including a fair rate of return on its investment. The exclusive right-to-serve protects the capital investment of the firm and the right to be served provides a similar protection for the customers' relation-specific investments by eliminating the customers' threat of switching to an alternative supplier or the threat by the firm of refusing to supply. As such, the entry barrier associated with exclusive rights-to-serve and rights to be served protect the investments of both consumers and suppliers and improve the incentives for investment.

The value of a right to serve for incumbent firms can be illustrated in a market where demand is growing. This is shown in Figure 3.2.

At time t demand is D_t, while at time $t+1$, demand has increased to D_{t+1}. The long-run average cost curve is denoted by

$AC(Q)$. Suppose that at time t, consumers and the firm agree that the firm will supply quantity Q_t at price P_t. When demand increases to D_{t+1}, consumers will have an incentive to elicit new entry. A new entrant would just be willing to provide Q_{t+1} units of output at price P_{t+1}. Of course this would strand the investment of the incumbent firm and it is likely to match the price and quantity of the entrant. However, it will not be able to do so for an average cost of P_{t+1} since its sunk capital investment is optimal for output level Q_t. Thus it will not recover all of its capital investment.[46]

The regulator can also play the role of a special court to adjudicate long-term contracts. A regulator may reduce the costs of monitoring, verification, and enforcement, as well as be a cost effective way of determining when adjustment to an unanticipated change is efficient. By doing so the regulator can provide considerable protection against the threat of hold-up and, as a consequence, increase the level of investment which the parties, both producers and pipelines, are willing to make. In essence, regulation can provide a more effective, less costly way to administer long-term relationships based on sunk investments than is possible under private contracting alone. For instance, producers are more likely to sign a contract for service where the toll is based on average cost if they know that a regulator will monitor and enforce the contract.

Requests for pipeline service that necessitate expansions typically require shippers to enter into long-term, open-ended commitments. These may involve contractual commitments to pay the demand charges for, say, 15 years. However, the contract often does not specify the amount of these demand charges: this amount is only determined over time as the regulator sets the tolls that can be charged. The difficulties that would arise if shippers and pipelines had to specify the terms of adjustment in a private long-term contract are not hard to imagine. In such circumstances, the parties are likely to make liberal use of arbitration clauses. This would amount to regulation by another name, with the arbitrator playing the role of the regulator. However, the costs associated with regulation may be less than the use of arbitration due to specialization and economies of scale.

Alternatively, in the absence of a regulator, shippers may be unwilling to commit to open-ended demand charges in long-term

46 This example is adapted from Sharkey (1982, 90).

contracts. This has two important ramifications. The first is that, presumably, pipelines would be reluctant to invest in new capacity. Second, the use of spot markets to allocate capacity may be incompatible with the present toll structure. Since straight fixed-variable (SFV) tolling has very attractive efficiency properties, this is an important issue which remains to be resolved.[47]

(iii) *The Role of Competition.* Some observers have argued that, since the ex-Alberta pipelines compete at the city gate with other supply basins, there is no strong justification for utility-type regulation for these systems.[48] However, this argument shows the importance of carefully defining the extent of the market. Even if there was in fact significant competition at the city gate, there could still be monopoly power out of a particular supply basin. The question is not only whether there is competition at the city gate in export markets, but whether there is sufficient competition to constrain the market power of the relatively few pipelines moving gas out of a particular supply basin.

Increased competition among supply basins may reduce the monopoly power of a single pipeline out of a supply basin, but it will not eliminate it. The existence of competition at the city gate will affect the elasticity of demand for transmission services connecting a supply basin and the city gate. Competition at the city gate between alternative supply basins decreases the elasticity of demand for bundled service from any one supply basin. This reduces the scope for, and the profitability of, monopoly pricing but does not eliminate the distortion.

There have been some recent studies which attempt to assess the state of competition that actually exists between natural gas pipelines (Gallick 1993; Kwaczek and Miles 1995). Gallick, using traditional measures of market structure to infer market power, concludes that in many instances in the U.S. competition at the city gate among existing pipelines and potential entrants is probably sufficient to eliminate the potential for significant monopoly pricing by pipelines.[49]

---

47 See Section 3.3 below for a discussion of SFV tolling.

48 For example, see the presentations to the workshop in 1992 on incentive regulation sponsored by the National Energy Board.

49 Gallick uses the Herfindahl index to measure market concentration. This index is the sum of the square of each firm's market share and has a value range from zero to one (the value is one for the case of a monopoly). Gallick implicitly assumes that high (low) concentration is correlated with high (low) market power.

Kwaczek and Miles use a similar approach to measure market power in pipeline transmission in Canada. They consider three separate markets: (i) the market for transmission out of Alberta (ii) the market for transmission between Alberta and Eastern Canada and (iii) the market for transmission into Eastern Canada. Their analysis suggests that all three markets are very concentrated. The number of options available to shippers in these markets is limited and deregulation is likely to result in the exercise of significant monopoly power.

Kwaczek and Miles adopt a perspective which is shared by others in the industry, namely that "… regulation substitutes for competition where it does not exist" (Kwaczek and Miles 1995, 1) and that if competition can be created, then it is desirable to do so. Given their finding that gas transmission in Canada is concentrated, they recommend that regulators follow policies which promote competition. These include easing entry restrictions, increasing flexibility in the terms of service, and removing restrictions on the establishment of a secondary market in capacity.

However, an implication of subadditivity and significant sunk costs is that easing entry restrictions and other measures which promote competition is not necessarily socially desirable. Promoting competition could result in higher industry costs and tolls, reduced investment by both producers and pipelines, and an inefficient toll structure.

(iv) *Competition for the Next Expansion.* The competition that is significant in Canada is the competition between corridors for the next expansion. Given the present regulatory regime where capacity expansions must be underwritten with long-term commitments by shippers, pipelines are in a position where they compete with each other for the next expansion. All other things equal, the pipeline which shippers perceive as providing the highest quality service at lowest cost will have an advantage. This provides pipelines with an incentive to build up a reputation for being cost minimizers.

There is some merit to this argument and, within a traditional cost of service regulatory framework, this competition clearly mitigates incentives for cost inefficiency. However, it is very different to argue that this competition justifies deregulation of pipeline transmission. The limitations associated with such arguments include the following.

(a) Transmission costs are only one factor which influences the profitability to shippers of an expansion. For instance, market conditions at the intended destinations of the gas will also play an important role. Pipelines in corridors where these factors are favorable may be able to exert significant monopoly power.

(b) Such an argument is not applicable to certain segments of the system. For example, they would not apply to the NOVA facilities since there are no similar (large, integrated) pipelines within Alberta competing for the business of carrying gas to export points at the provincial boundary or to many intra-Alberta markets.

(c) The argument is, to a certain extent, circular. The existence of competition for the expansion arises because of the current regulatory regime which typically requires long-term contracts for regulatory approval of the expansion. In the absence of this regulation, such contracts would not be required.

(d) Some pipelines might find it more profitable to not compete for expansions, but to exercise a degree of monopoly pricing in existing facilities.

Proponents of the view that competition for expansions is sufficient to justify deregulation have yet to demonstrate that it would in fact result in socially desirable outcomes given the subadditivity of costs and the extent of sunk expenditures. The presence of these characteristics would lead to the conclusion that competition in and of itself is not necessarily socially desirable.

Inequity. An outcome may be socially unacceptable on equity grounds, even if it is efficient. In such circumstances, governments may intervene and attempt to alter the market outcome because of objections related to the unfairness of the distribution of benefits in an unregulated market.

In the case of public utilities, equity considerations are typically limited to the concepts such as 'just and reasonable' tolls, 'no undue discrimination', and 'cost-based tolls.' In cases involving pipelines, regulatory tribunals have not tended to set prices or tolls with the primary objective of changing the distribution of income or resources toward one that is 'more fair.' First, fairness in this broad sense is very much dependent on value judgements. Second, since they are not elected, regulators are typically not in a strong position to exercise value judgements. This is usually the role of elected representatives who have the responsibility for achieving a 'fair' or 'equitable' distribution of income, as defined by the electors, through the taxation and government expenditure system. Finally, it is generally recognized that the first objective must be efficiency. Unless efficiency is achieved, redistribution issues are rather academic because there will be little to redistribute.

Responses to Market Failure. The first step in determining whether regulation is appropriate involves determining whether there are significant elements of market failure and resulting inefficiency.[50] The second step involves establishing the proper nature of the intervention. Given existing informational, institutional, political, and technological constraints, a determination must be made as to whether it is possible to intervene in a manner which corrects the market failure or inefficiencies. The third step involves establishing whether the benefits from intervening justify the costs of intervention. The applicable regulatory costs do not just include the direct costs associated with the administrative agency and its procedures. They also include costs associated with any induced misallocation of resources. It may well be that, in some cases, regulation in practice leads to a reduction in societal welfare rather than an increase. To put it differently, it is important to establish that the outcome with costly and necessarily imperfect regulation is a significant improvement over the outcome without regulation even if there are significant elements of market failure.

If the markets for gas transmission were not regulated, the resulting market structure might be quite concentrated since the existence of significant economies of scale is inconsistent with workable competition. In a competitive market, prices are driven down to the level of marginal cost. However, when there are economies of scale, average cost exceeds marginal cost and, consequently, there will be negative economic profits if the firm prices at marginal cost. Negative economic profits mean that firms will be earning below-average profits and will exit the industry. They will continue to do so until the firm(s) remaining have sufficient market power to raise the price level to at least that of average cost. Thus, the existence of economies of scale will limit the number of firms that can survive.

This unregulated market would be inefficient due to monopoly (or oligopoly) pricing. Moreover, if it entails service provision by more than a single firm, total industry costs will be higher than if service was provided by a single firm. Finally, given the importance of sunk costs, investment levels could easily be inefficient due to the hold-up problem

50 This version of the market failure test follows Spulber (1989, 3).

and the associated costs of private contracting, and unregulated markets may not be able to sustain long-term contracts with SFV toll structures. The costs of these inefficiencies must be compared to the costs and distortions associated with regulation. It is shown in the next section that the costs associated with monopoly pricing alone can be significant.

Estimates of Losses Due to Market Power. It does not appear that there has been a comprehensive and detailed analysis to quantify the benefits and costs of regulation of the major gas pipelines. Following Waterson (1988) and Schmalensee (1982), it is instructive to make some 'ball-park' calculations of the social losses that would arise if a major monopoly pipeline were allowed to set prices without regulatory oversight.

These calculations are illustrated in Figure 3.3. Suppose that the demand curve for gas transmission can be approximated by the linear curve $P(Q)$; the corresponding marginal revenue curve is $MR(Q)$. Let marginal costs be constant and equal to c. The profit maximizing monopoly output is Q^m, where marginal revenue equals marginal cost. The output which maximizes total surplus is Q^s. The shaded area represents the lost gains from trade (total surplus) associated with monopoly pricing.

This area is equal to the area of a triangle with base $\left(Q^s - Q^m\right)$ and height $\left(P^m - c\right)$. It can be shown that this equals $\frac{\left(P^m - c\right)}{2P^m} * \left(P^m Q^m\right)$.[51] The first term is half of the price-cost margin ($\frac{\left(P^m - c\right)}{P^m}$) and the second term ($P^m Q^m$) is total revenue. If prices were marked up over cost by 50 percent, then the price-cost margin would equal a third. If total revenues were $1.5 billion annually, the social loss associated only with monopoly power would equal $250 million annually. The potential gains from reducing the distortion associated with monopoly pricing can be quite significant. The important point here is that while regulatory costs can sometimes be quite large, the costs to society of not regulating when workable competition is not possible can be much larger than these regulatory costs.

51 This equation holds only for the case of linear demand and constant marginal costs. See Schmalensee (1982, 1809) for the derivation.

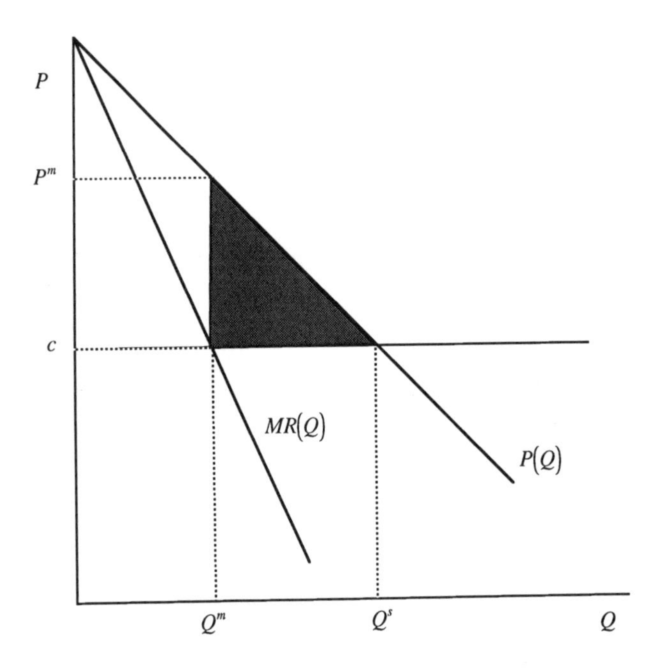

FIG. 3.3 EXAMPLE OF WELFARE LOSSES FROM MONOPOLY PRICING

Also, this may partly explain the existence of differences in the degree (and cost) of regulation based on differences in the size of the firm. For example, the more extensive (and expensive) regulation applied by the NEB to Group 1 oil and gas pipelines compared to that for the Group 2 pipelines may be justified by the fact that the social losses that would exist in the absence of regulation for the larger pipelines could be quite large.

3.3 OPTIMAL PRICING

Once the decision to regulate prices or tariffs has been made, the question that arises is how they should be set to achieve efficient outcomes: that is, "What are the socially optimal prices?" There are two aspects of optimal pricing: the first concerns the overall price level, while the second concerns the structure of relative prices.

The issue of optimal relative prices arises when a firm produces more than one product. A relative price is simply a price ratio. An example in the context of pipelines would be the toll for firm service to point A relative to the toll for interruptible service to the same point.

Single Product Firms. Consider the problem faced by a regulator who has the opportunity and power to set the price for a single product natural monopoly. If the focus is on an efficient solution, the regulator will try to set a price and output combination which maximizes total surplus. This unconstrained problem replicates the competitive pricing rule: set price equal to marginal cost. Figure 3.4 illustrates that 'first-best' efficiency is achieved by setting price equal to marginal cost and producing Q^S units of output. However, if the firm produced this level of output it would incur a loss of $AC(Q^S) - P^S$ per unit. A regulator would have to provide an equivalent subsidy or the firm would refuse to invest. Since regulators do not typically have the power to levy taxes and pay subsidies, a 'second-best' approach must be used.

The problem facing a regulator is to choose a price and quantity which maximizes total surplus subject to a break-even constraint for the firm: that is, the firm is allowed to earn a normal return or level of profit.

Prices which solve this problem are known as Ramsey prices. In Figure 3.4, the Ramsey price and quantity pair is (Q^R, P^R). The difference in total surplus between the first and second-best outcomes is the shaded triangular region. Total surplus is smaller since prices must be raised above marginal costs in order to fund the operations of the firm. The total reduction in consumer surplus is the sum of the shaded and striped regions. The striped region is the reduction in consumer surplus which is transferred to the firm so that it breaks even. The second-best solution involves determining how to most efficiently transfer surplus (as revenues less variable costs) to the firm so that it is financially viable (that is, so revenues less variable costs equal fixed costs).

Multiproduct Firms. It will be recalled that most pipelines are really multiproduct firms since the various types of service provided constitute different commodities. The first-best pricing for a firm which produces more than one product or service involves setting the price of each commodity equal to marginal cost. However, if the firm is a natural monopoly, this means that while such prices maximize total surplus, the firm will incur losses. Unless the regulator can pay a subsidy, the next-best approach will be to once again use Ramsey pricing.

The second-best efficient solution is complicated by the existence of common costs. Common costs are those which cannot be attributed to the provision of any particular product. These costs are usually the source of economies of scope. The second-best solution involves deter-

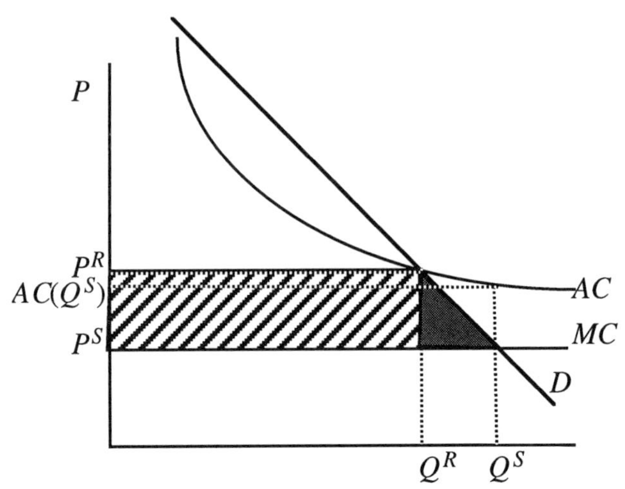

FIG. 3.4 FIRST AND SECOND BEST PRICING

mining how to meet the revenue requirement of the firm and, at the same time, minimize the loss in total surplus from raising prices above marginal cost.

Ramsey prices for a multiproduct firm follow the inverse elasticity rule: prices are set such that their deviation from marginal costs is in inverse proportion to the elasticity of demand. In markets where the price elasticity of demand is high or, alternatively, where demand is very price-sensitive, prices are set close to marginal cost. Raising prices in such markets significantly reduces the quantity demanded and, therefore, reduces total surplus. In markets where the price elasticity of demand is low or, alternatively, where demand is not very price-sensitive, raising prices does not have a large effect on demand. Consequently, in these markets the loss in total surplus from raising prices is small.

Formally, the Ramsey pricing rule when there are two goods with independent demands is

$$\left(\frac{P_1 - MC_1}{P_1}\right)\varepsilon_1 = \left(\frac{P_2 - MC_2}{P_2}\right)\varepsilon_2,$$

where ε_1 and ε_2 are the price elasticities of demand for commodity 1 and 2, respectively.[52] P_1 and P_2 are the prices for these commodities and MC_1 and MC_2 are the respective marginal costs.

52 For instance $\varepsilon_1 = \frac{\%\Delta Q_1}{\%\Delta P_1}$, the percentage change in quantity demanded divided by the percentage change in price.

An intuitive explanation for the Ramsey pricing rule is presented in Figure 3.5.[53] Consider the two good case. Let commodity 1 be relatively price-elastic and commodity 2 be relatively price-inelastic. Suppose that both commodities have the same marginal cost of production, $MC_1 = MC_2$. Suppose that at the first-best prices, demand for the two products would be the same, $Q_1^s = Q_2^s$. Consider the effect of raising the price by the same amount in each market, to $\tilde{P}_1 = \tilde{P}_2$ so as to cover fixed and variable costs for the regulated firm. Raising price above marginal cost has two effects. It transfers consumer surplus to producers (the striped regions) and there is a loss in surplus due to the reduction in quantity (the shaded triangles). The same increase in price has very different effects. In the elastic market, the price increase results in a large reduction in quantity and reduction in surplus, relative to the situation in the inelastic market. Moreover, the transfer of surplus to the firm is less in the elastic market relative to the inelastic market. In effect the transfer in the elastic market is quite costly in terms of lost surplus relative to the inelastic market and, as a consequence, prices in elastic markets should not be increased (relative to marginal cost) as much as prices in inelastic markets.

It should be noted that the same principles apply when there is one commodity which is sold in two or more distinct and separable markets. For example, Ramsey pricing would involve setting prices closer to marginal cost in the elastic market and farther from marginal cost in the inelastic market.

Applicability of Ramsey Pricing. Four features of Ramsey pricing limit its applicability. The first is that the information requirements to implement such pricing are formidable. Detailed information about demand and cost functions is required. Nevertheless, it may be possible to at least move in the direction of Ramsey pricing through the use of qualitative or judgmental estimates of the demand elasticities.

The second difficulty is that moving from existing prices to Ramsey prices will likely create winners and losers. Consumers with elastic demands will be better off because the prices they pay will typically fall.[54] Consumers with inelastic demands will be made worse off by a move to Ramsey prices because the prices they pay will usually rise. Ramsey prices are efficient because they maximize total surplus, the dollar measure of gains from trade. This means that the dollar value of

---

53 Figure 3.5 is adapted from Train (1991, 122–123). See Berg and Tschirhart (1988) for a more formal presentation of Ramsey pricing.

54 This is true provided the existing price for the elastic market is above marginal cost.

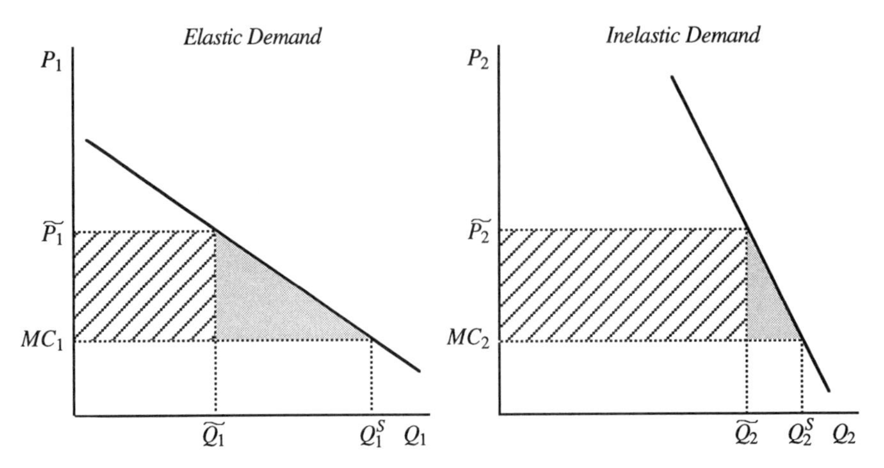

FIG. 3.5 RAMSEY PRICING: THE TWO GOOD CASE

the gains of the winners will exceed the dollar value of the losses of the losers. However, unless compensation is paid, there will be conflict associated with a move to Ramsey pricing.

The third feature is that prices under such a system are not strictly cost-based. Prices are a function of both marginal cost and the elasticity of demand. Finally, Ramsey pricing is essentially price discrimination on the basis of willingness to pay. As such, it may run afoul of legislation requiring tolls to be 'just and reasonable.' With regard to the last two issues it might be noted that legal arguments have been made on a number of occasions that Section 62 of the NEB Act (which refers to just and reasonable tolls) mandates cost-based tolls.

Inefficient Entry and Cross-Subsidies. The last two features imply that Ramsey prices are not *necessarily* subsidy-free, though they may well be. In some cases there will be an incentive for inefficient entry or inefficient by-pass. Prices which are not subsidy-free, when economic profits are zero (or the level of accounting profits is in a 'normal' range), imply that consumers of one product are paying revenues in excess of the stand-alone costs of providing services. This provides an opportunity for profitable entry.

The following example from Train (1991, 311–313) illustrates this possibility for a two good case. Suppose that demand for good 1 is perfectly inelastic: regardless of the price, 1000 units are demanded. Good 2 is price-elastic with the following demand function: $Q_2 = 1280 - 10P_2$. The *stand-alone* cost functions for good 1 and good 2 are $C_1(q_1) = 20{,}000 + 2q_1$ and $C_2(q_2) = 30{,}000 + 3q_2$. The cost function for a firm which produces both products is $C(q_1, q_2) = 40{,}000 + 2q_1 + 3q_2$. Notice that the cost function is subadditive: joint production of any amount of either good reduces total costs by $10,000 compared to producing the same amount of each good separately.

Since demand for good 1 is perfectly inelastic, Ramsey prices are simply $P_2^R = 3$ and $P_1^R = 42$. The Ramsey rule requires that good 2 be priced equal to marginal cost and all of the fixed costs are funded by raising the price of good 1 alone above marginal cost. This result obtains since raising prices in the market for good 2 would create a large distortion in the quantity of good 2 in the market, but raising the price for good 1 does not create such a distortion in the market for that good.

However, it can be noted that these prices involve a cross-subsidy. Revenues from the first market are $42,000 but stand-alone costs are only $22,000 and conversely, revenues in the second market are $3,750 while stand-alone costs are $33,750. A potential entrant could profitably enter and supply only market 1 by charging a price between $22 and $42. This entry would be privately beneficial for the entrant and consumers in the first market. However, their gains would not make up for the losses consumers of good 2 would incur. The social inefficiency of entry would be further exacerbated if the incumbent's fixed costs were sunk.

Non-Linear Pricing. Implicit in the above discussion of Ramsey pricing was the restriction that the firm could only charge a linear price.[55] In the context of pipelines, a linear price is one where there is only a commodity toll so that the revenue is simply this price times the volume shipped. That is, there is no 'demand' component to the tolls charged.

Coase (1946) was the first to observe that non-linear pricing may be a more efficient way to meet the revenue requirement of the firm. A common form of non-linear pricing is a two-part tariff where the total revenue equals a fixed fee which is independent of actual commodity sales, and a price per unit of the service actually used. The demand / volumetric tariff structure of natural gas pipeline tolls is an example of a two-part tariff. The demand charge or access fee is a means whereby consumers can pay the amount required to fund marginal cost (first-

---

55 Uniform prices (p) are linear, since the total bill of the consumer (p*q) is linear in consumption, q.

best) pricing.[56] In general, this will introduce fewer distortions than if the revenue deficiency was to be recovered only by raising the variable or commodity charge.

3.4 REGULATORY OBJECTIVES AND CRITERIA

Regulation replaces the invisible hand of the market with the visible hand of the regulator. For regulation to be effective, the regulator must induce the firm to act in a socially desirable manner. This will happen when the incentives and constraints the firm faces are such that profit maximization by the firm leads it to the socially desirable outcome. The two tasks of regulation are to identify socially desirable outcomes and to institute a regulatory mechanism under which the socially desirable outcome will be realized when the firm maximizes its profits. There are a number of objectives and characteristics of an ideal regulatory framework for a natural monopoly. It is important to note, however, that no regulatory regime will be ideal. Instead, each will have its own strengths and weaknesses and any decision regarding which one to implement will require an assessment of the relative merits of each alternative.

Regulatory Regime Objectives. The six standard objectives of an ideal regulatory regime or constraint are as follows.

(i) *Efficiency*. The regulatory constraint should induce the firm: to produce at least-cost (cost efficiency); to supply an efficient output level and to distribute those supplies efficiently; to adopt cost-reducing technologies and introduce new products or services whenever they can be justified on a cost-benefit basis; and to supply a quality of service (for example, reliability of service) and types of service that are socially optimal in the sense that the benefit of an increment of quality is equal to the cost of that increment.

56 Two-part tariffs can be interpreted as involving paying a price for access, the fixed component, and a price for usage, the price per unit actually consumed. If access demand is perfectly inelastic, then following the Ramsey pricing rule, only the price of access should be raised above marginal cost and the usage component of the two-part tariff can be set equal to marginal cost. If the price of access is set too high, some consumers may find that the benefits of consumption at a usage price equal to marginal cost are less than the costs of access and, hence, they will not purchase access. In this case access will be price elastic and the Ramsey solution will involve raising both the access price and the usage price above marginal cost.

(ii) *Cost Reductions Benefit Consumers.* An ideal regulatory structure ensures that consumers (shippers in the case of pipelines) also benefit from reductions in cost.

(iii) *Economic Viability/Sustainability.* The structure should provide sufficient revenues so that the regulated firm remains viable. Also, the regulatory regime should be sustainable over a lengthy period of time.

(iv) *Fairness/Equity.* An ideal governance structure recognizes fairness concerns. The procedures followed and the outcomes achieved should be perceived as fair. The precise meaning of fairness/equity is often unclear. Nevertheless, as noted in the next section, there are a number of traditional elements used by many regulatory agencies which form a workable definition.

(v) *Regulatory Burden.* The costs incurred by all participants and interested parties should be minimized. The costs from participating in the regulatory process should not be onerous and the direct administrative costs should be minimized.

(vi) *Implementability.* The regulatory rules and processes should use readily available, easily measured, and verifiable data. Also, they must be practical from an administrative viewpoint and they should have general acceptance.

Traditional Regulatory Criteria. The preceding dealt with criteria for evaluating regulatory regimes. However, some parts of the subsequent analysis involve the use of the decision criteria commonly used by regulators. These criteria are discussed below.

The regulatory regimes associated with agencies such as the National Energy Board (NEB) and provincial Public Utility Boards (PUBs) have tended to embody a fairly standard set of specific criteria for making tolling and related regulatory decisions. In some cases these criteria are set out in the legislation defining their mandate; in other cases, they are based on precedent and general acceptance. The most important of these criteria are summarized below.

(i) *Fairness and Equity.* This usually encompasses several elements. These are the requirements that tolls be 'just and reasonable' and that they not constitute 'unjust discrimination'. Such terms are rarely precisely defined but generally embody the notions that tolls should be equal for equal service, that tolls should be based on cost causality, that tolls should be consistent with the principle of 'no acquired rights' and that tolls should not unjustly discriminate against any party. Cost causality refers to the principle that the tolls should reflect those costs caused by the provision of service to the particular customers. The principle of no acquired rights means that customers do not acquire ownership or other special rights to the facilities used to provide service simply because they have contracted for and used those services in the past.

(ii) *Rate Stability.* Again, this is typically not defined with any precision. It generally means that rates should be reasonably predictable and that there should not be 'rate shock'. As a rule of thumb, rate shock refers to situations where annual rates of increase in tolls are in 'double digits'. However, the context is also important. For example, a significant increase in tolls would be viewed differently if real tolls had been declining and were expected to continue to decline after the spike than if real tolls had been rising and were expected to continue to rise. This criterion also embodies the notion that there should generally be a gradual transition to new rates to avoid hardships on particular customer groups.

(iii) *Encouragement of Efficiency.* This involves the definitions of efficiency as outlined earlier.

(iv) *Revenue Sufficiency and Stability.* This refers to the requirement that the tolls provide adequate revenues to meet all necessary costs and provide a fair return to investors, while maintaining appropriate service and safety levels. It also concerns the desirability of a reasonably predictable revenue stream.

(v) *Consistency With Other Policies and Regulation.* This mainly concerns the consistency of regulatory decisions with the objectives of natural gas market and price deregulation and with policies such as the North American Free Trade Agreement (NAFTA). With regard to deregulation, it is particularly important that the tolls provide the proper market signals and efficiency incentives so that the deregulated energy markets operate efficiently.

(vi) *Practicality, Administrative Simplicity and General Acceptance.* This generally means that the tolling methodology should be well-understood, that the methods used to set the tolls should be as logical and as straightforward as possible, and that the tolls and methodology should be as free as possible from controversy.

4 Traditional Regulatory Regimes

The most common regulatory regime is that referred to as Cost of Service (COS) regulation. This traditional approach involves the regulation of prices or tolls so as to cover all prudently incurred costs in providing the product or service, including a 'fair' return on investment. There are, however, a number of variations. For example, it could involve 'light-handed' regulation, such as when regulation is on a complaints basis rather than a more active approach to regulation. Other variations include streamlined regulatory processes and the incorporation of particular incentive measures.

The main elements of traditional COS regulation are outlined in Section 4.1. Other variations are discussed in Section 4.2. The tolling patterns associated with COS regulation are outlined in Section 4.3. Section 4.4 provides an overall evaluation of this traditional approach to regulation in terms of the concepts, objectives and other background details dealt with in Sections 2 and 3.

4.1 TRADITIONAL COS REGULATION

Rate of Return (ROR) Regulation. Cost-of-Service (COS) regulation is an extension of Rate of Return (ROR) regulation. Under the ROR approach, the maximum rate of return on the firm's capital is set to reflect a value deemed to be 'fair' by the regulator. The firm is then allowed to set prices, choose inputs and make other cost and revenue decisions so long as the actual rate of return does not exceed the set fair rate of return.

This approach can be conveniently summarized using the following notation.

$$f \geq r \text{ (or, alternatively, } r \leq f) \qquad (4.1.1)$$

where f is the 'fair' rate of return and r is the actual rate of return, and

$$r = (P*Q - w*L - C) / K \qquad (4.1.2)$$

where P is the unit price / toll, Q is the quantity or volume of service, w is the unit cost of labour inputs, L is the number of units of labour used, C represents other input costs, and K is capital or, in this context, the rate base. K would be calculated as gross plant (GP) minus accumulated depreciation (D) and contributions in aid of capital ($CIAK$) or 'surcharges,' plus working capital (WK) and any net capital deferrals (KD).

Equation (4.1.1) simply states that the actual rate of return must be less or equal to the set rate of return. According to the second equation, the rate of return is total revenue ($P*Q$) less labour ($w*L$) and other (C) costs, all as a percentage of total capital or the rate base (K). Under strict rate of return regulation, the regulator only chooses f and the firm selects the price / toll level (P), the amount of labour (L), perhaps to some degree the wage and benefit level (w), the level of other non-capital costs (C) and the level of the rate base (K). The rate base will be determined by decisions concerning capacity additions, depreciation rates and methodology, capital contributions by shippers and levels for working capital. The values for all of these variables must be selected such that the combination, given the actual volume or quantity of service (Q), does not violate the rate of return constraint (Equation 4.1.1).

This type of regulatory regime has been assumed in much of the literature dealing with such things as the Averch-Johnson effect where the regulated firm, it is alleged, will be inefficient because of perverse incentives.[57] However, as emphasized by Joskow (1974), in most cases of public utility regulation the firm is not allowed the freedom assumed in the Rate of Return (ROR) regulation model. That is, COS regulation is more commonly used and under that regime most of the other variables (such as P, w, L, r, and K) will either be set by the regulator or the level of each will be heavily influenced by the regulator. This is very different from the process envisioned under a strict ROR regulation scheme and, consequently, many of the inefficiencies and shortcomings emphasized in the literature (which are based on a strict ROR regulation regime) may not be applicable to the schemes typically employed in practice.

57 This effect is often described as the alleged tendency for regulated firms under rate of return regulation to inflate the size of the rate base if the fair rate of return set by the regulator exceeds the firm's cost of capital. However, it is more accurately described as the incentive for regulated firms to use an inefficiently high capital-labour ratio when the fair rate of return set by the regulator exceeds the firm's cost of capital. See Averch and Johnson (1962).

Traditional Cost of Service (COS) Regulation. This approach involves much more involvement by the regulator who will make or approve decisions concerning every variable indicated in the simple model outlined above. In general, it will involve a sequential determination of the following elements.

(i) *Allowed Rate Base*. This requires decisions on gross plant carried from previous years, new additions, depreciation, capital in aid of construction (or surcharges), working capital, and various deferrals.

(ii) *Allowed Cost of Capital*. This requires a determination of the cost of funded debt, the cost of unfunded debt and the appropriate return on equity.[58] To simplify, suppose that the cost of debt is r_d and the return on equity is r_e. Further, the regulator will typically rule on the appropriate proportions of debt and equity financing. If these deemed proportions are denoted 'd' and 'e', the approved cost of capital (denoted 'r') will be the weighted average cost of debt and equity given by the equation: $r = (d * r_d) + (e * r_e)$.

(iii) *Allowed Return on Rate Base.* This simply requires multiplication of the approved rate base (K) by the approved cost of capital (r).

(iv) *Allowed Depreciation Expense.* This requires a decision to be made on the appropriate rate of depreciation to be applied to the various types of capital. A common method of setting this rate is Average Service Life but other methods exist. In any event, this rate is usually applied using straight-line depreciation methodology.

---

58 There are three main approaches used in determining the appropriate return on equity. The Comparable Earnings Test is based on the opportunity cost principle that the regulated utility should be allowed to earn returns on common equity that are comparable to the returns on common equity earned by non-regulated companies that have comparable levels of risk. The Discounted Cash Flow (DCF) Test attempts to measure the investors' expected return on common equity by the current share prices in the market and the estimated future dividend cash flow. It assumes that the market price of a share is equal to the future cash flows associated with the shares, discounted at a rate which reflects the investors' required return for a particular risk level. The Equity Risk Premium Test is based on the notion that there is a close relationship between the level of risk and the rate of return required to induce investors to hold a particular instrument. This risk premium is usually established in relation to low-risk instruments such as long-term government bonds. In recent years, this test has generally been given the most weight in determining the appropriate return on equity for regulated utilities. As noted in Section 4.2, the NEB recently introduced a formula approach in determining the return on equity for a number of pipelines. It involves setting a return on equity for a benchmark pipeline, primarily using the equity risk premium technique, and then adjusting this over time by some fraction of the change in long-term Government of Canada bond rates.

As noted in Section 4.3, this standard approach to depreciation results in 'front-end loading' whereby, all other factors constant, tolls will decline over time.

(v) *Allowed Operations, Maintenance and Administrative Expenses.* This typically will require the regulator to review such things as the number of employees, the level of wages and benefits, the amount and cost of office space, the amount and cost of maintenance work and so on.

(vi) *Taxes Payable.* This involves the estimation of municipal, capital, income and other taxes which will be levied on the firm and a decision (which is usually for flow-through taxation) on how any tax deferrals are to be treated in the calculation of current costs and revenue requirements.

(vii) *Other Attributable Costs.* The costs associated with a variety of items must also be estimated. These would include items such as costs for services purchased from other connected pipelines, regulatory costs, interim toll adjustments and often a number of deferrals.

(viii) *Revenue Requirement.* The revenue requirement (*RR*) will be an estimate of the total revenue that must be collected in order to cover the total costs computed by adding the costs in (iii) to (vii) above.

(ix) *Allocation of the Revenue Requirement or Total Cost.* Given that multi-part tolling is typically used, it will usually be necessary to separate the *RR* or total cost into those components which are fixed (that is, do not vary with throughput) and those which are variable. Under the straight fixed / variable (SFV) methodology, the former costs would be recovered under the 'demand' or 'access' part of the toll and the variable costs would be recovered under the 'commodity' part of the toll.[59] Further, costs must be allocated by type of service and, unless postage stamp tolls are used, it will also be necessary to allocate this *RR* or total cost on a zone basis. The usual way of doing this involves using a mixture of indices based on such things as distance and volume. This

59 Other approaches, such as Modified Fixed / Variable methodology, have been used in the U.S. where the purpose has been to alter the allocation of risk associated with under-utilization. Such approaches make the pipeline shareholders, rather than just the shippers, bear some of the cost-recovery risks.

approach of allocating costs to determine prices is referred to as Fully Distributed Cost (FDC) pricing.

(x) *Setting the Tolls or Prices.* The last step involves translating these allocated costs into tolls. This is usually accomplished by determining the appropriate quantity (Q) of each service which will be used to establish the unit price or toll for each service. Depending on the scheme, it could be actual throughput, expected throughput or some volume differing from expected throughput in order to redistribute the risks associated with under-utilization of capacity. In some cases, such as the methodology used by the NEB, the price is set equal to the allocated cost divided by the appropriate volume. In other cases, for example the approach used by the Alberta PUB in setting electrical rates, there is a requirement that the tolls or prices for various segments or services not deviate from the allocated per unit costs for those segments or services by more than some fixed percentage, such as 5 percent.

To summarize, using the simple, single product example outlined earlier, traditional COS regulation involves first the setting of the Revenue Requirement (RR) equal to the sum of the approved costs $r^*K + w^*L + C$ (where C represents the costs other than those attributable to capital or labour). This RR is then divided by a value for total quantity or volume (Q) to arrive at the regulated price or toll (P). That is, the regulator sets the price as:

$$P = (r^*K + w^*L + C) / Q \qquad (4.1.3)$$

It is also useful to highlight the 'regulatory compact' that has evolved in most cases of COS regulation in Canada. In most cases of natural monopoly, the regulated firm has, either formally or informally, been granted a franchise giving it the right to be the sole supplier and the obligation to serve the complete market. In return, the firm accepts that it will be regulated to serve the public interest and the regulator accepts that it will allow the firm the opportunity to earn a reasonable return on its investments. The latter means, for example, that the shareholders are to be protected from factors such as changing interest rates or throughput variations which are beyond the control of the firm or which cannot often be accurately forecast at the time decisions are made. One manifestation is the large number of deferral accounts designed to ensure that the shareholders are allowed full recovery of prudently incurred costs even when they can only be accurately determined well after the regulatory tolling and other decisions have been made.

4.2 OTHER VARIATIONS

While the approach as described above is the most common version of COS regulation, there are numerous variations that can also be observed. These are summarized below.

Active versus Light-Handed Regulation. In cases such as the regulation of the Group 1 gas pipelines by the National Energy Board (NEB), the regulator plays a very active regulatory role.[60] In addition to the requirement of detailed reporting by the firm, the regulator continuously monitors the firm's operating results, regularly conducts audits and holds frequent hearings to establish tolls and implement other constraints.

An alternative is a light-handed approach where, for example, regulation is on a complaints basis. That is, unless a complaint is brought before the tribunal, the regulator does not require public hearings or other costly forms of intervention. However, in the event of a complaint, the more active approach is initiated. An example of the light-handed approach is that which has been used until recently in the regulation of the gas transmission component of NOVA (as of January 1, 1995, this component of NOVA became more actively regulated by the AEUB). Another example of light-handed regulation is the approach used by the NEB in the regulation of Group 2 pipelines.

Ex Post versus Prospective Rate Making. One approach under COS regulation is to essentially allow the firm to recover its costs on an *ex post* basis. For example, the tolls will vary according to the monthly costs actually incurred by the regulated firm and by actual variations in flows. This is the variant applied by the NEB to Foothills Pipe Lines Limited (Foothills), for example.

The more common approach is to set tolls on a prospective or 'test year' basis. Here, estimates are made for all of the cost items for a future year and these forecasts, as well as those for throughput, are then used to establish the approved tolls. This means that only if all of these forecasts were realized would the firm be assured of earning the approved rate of return. As previously noted, some deviations from the

---

60 In the case of gas pipelines, Group 1 includes Alberta Natural Gas Company Ltd., Foothills Pipe Lines Ltd., TransCanada PipeLines Limited, Trans Quebec and Maritimes Pipeline Inc., and Westcoast Energy Inc. In the case of oil and products pipelines, Group 1 includes Cochin Pipe Lines Ltd., Interprovincial Pipe Line Inc., Interprovincial Pipe Line (NW) Ltd., Trans-Northern Pipelines Inc., and Trans Mountain Pipe Line Company Ltd. All other pipelines under NEB jurisdiction are classified as Group 2.

forecast values will often be recoverable later via deferral accounts provided that the deviations were not under the control of the firm or were not reasonably foreseeable. However, other deviations will affect the rate of return realized by the firm.

The implicit 'regulatory lag' is an important facet of the 'test year' variant of COS regulation. Once tolls are set, the firm has the opportunity to reduce controllable costs and, if successful, it would be allowed to keep the resulting higher profits until the next regulatory hearing. This feature is frequently mentioned in the literature as a significant incentive for such regulated firms to increase cost efficiency.

Two-Part Tolls. The most common tolling methodology involves two-part tolls consisting of a demand or access fee and a variable or commodity toll. In Canada, the usual approach has been to allocate the costs to these two components using straight fixed / variable methodology. That is, the pipeline company's fixed costs are fully recovered through the demand charge while its costs that depend on throughput are recovered through the commodity charge. This has the effect of greatly reducing the risk to the pipeline associated with load factor variations. Further, as noted in Section 3.3, these two-part tolls have important efficiency advantages over single-part tolling schemes.

There are other methodologies that can be and have been used with COS regulation. For example, modified fixed / variable tolling has frequently been used in the U.S. Under that approach, some of the fixed costs are recovered through the commodity component of the toll. As such, the pipeline is put at risk for recovery of some of the costs if high load factors are not achieved. The intent is to provide an incentive for the pipeline company to behave in a manner that leads to the maximization of throughput. This type of approach made more sense in the era of merchant pipelines where the pipeline companies had much more control over the factors that affected total throughput. As contract carriers, they have only minor control via incentive tolling schemes or new, more flexible services, assuming that these would be approved by the regulator.

Streamlining. There are a number of changes that can be made to traditional COS regulation which are primarily aimed at streamlining the regulatory process, cutting regulatory costs and increasing the effectiveness of regulation. One of these is the use of committees, task forces and negotiated settlements to reach agreements outside of a formal or public hearing process. Only in the event of a failure to reach a consensus or settlement would the usual, formal regulatory process be initiated. Other potential modifications to streamline COS regulation include the use of formulas to set allowable rates of return on equity (for example, setting each pipelines allowed rate as some fixed amount above the yields on long-term government bonds), the use of generic hearings to set a common rate of return on equity rather than having to consider rate of return issues for each pipeline at separate hearings, and the exemption of certain routine investments or investments in previously approved multi-year plans from the formal hearing process.

The NEB has expressed interest in modifications to streamline COS regulation and has recently implemented a number of such modifications. For example, in addition to encouraging negotiated settlements and exempting certain routine or multi-year investments from formal hearings, it has moved to a generic, formula approach for determining the cost of capital for six major pipelines.[61] With respect to the latter, the Board has established an appropriate rate of return on equity for a benchmark pipeline to serve as the standard in setting this return for six pipelines. It decided that the all-inclusive equity risk premium for the benchmark pipeline would be 300 basis points so that if, for example, the long-term bond rate was 9.25 percent, the appropriate rate of return on common equity would be 12.25 percent. This was the rate of return selected for the 1995 test year, the base year in applying subsequent adjustments.

The Board argued that differences in risk among the six pipelines should primarily be taken into account via the deemed debt-equity ratio. After considering the various risks, it decided that the deemed debt-equity ratio should be 30 percent for ANG, Foothills, TransCanada and TQM, 35 percent for Westcoast and 45 percent for Trans Mountain Pipe Line Company Ltd. (TMPL).

The last component of this streamlining initiative is a formula to adjust the equity return at the beginning of each subsequent year, thereby avoiding the costly hearing process. This involves five steps. First, a forecast of long-term Government of Canada bond yields is derived from Consensus Forecasts, a publication of Consensus Economics Ltd. of London, England. Second, this is subtracted from the test year bond yield used for the preceding year and the difference is multiplied by a factor of 0.75. This product is the adjustment to the rate of return on common equity for the test year. That is, for every change of 100 basis points in expected long-term Government of Canada bond yields, the allowed return on common equity changes by 75 basis points. Third, this product is then added to the rate of return on

61 See NEB (1995b).

common equity in the preceding test year. Fourth, the result is rounded to the nearest 25 basis points. Fifth, each pipeline subject to this streamlining then files a revised tariff of tolls for the current year which incorporates all of the elements outlined above.

Incentive Modifications. A number of the incentive mechanisms receiving increasing attention in recent years can be grafted onto traditional COS regulation. This includes schemes such as banded rates of return, profit sharing, performance benchmarking, and capital cost incentives. In fact, some incentive variations are already employed in Canada.[62] The various incentive alternatives are discussed in later sections.

4.3 TOLL PATTERNS UNDER COS REGULATION

Front-End Loading. The straight-line approach is commonly used to calculate depreciation for the purpose of setting tolls under COS regulation. As noted in Navarro et al. (1981) and illustrated in Figure 4.1, this depreciation methodology leads to a strong front-end loading pattern in tolls. That is, in the absence of capacity additions or changes in interest rates or rates of inflation, the tolls are high in the early years and in the later years, after the facilities are heavily depreciated, they fall to levels far below marginal cost.

In practice, the temporal toll pattern will be less regular. An example, as shown in Figure 4.2, is the pattern of actual and projected tolls for the Eastern Zone on the TCPL system. In real terms, the tolls are fairly constant. The pattern shown is a consequence of the interactions of three main factors. First, there is a fundamental downward trend in tolls during periods of stable capacity as the rate base declines through depreciation. Second, this is modified by the changes in inflation rates which result in operating cost variations and changes in the cost of debt and equity. Finally, there are short-run increases following significant expansions primarily because of the additions to the rate base.

Efficiency and Proper Price Signals. For a 'balanced' expansion involving compression and looping, the real, long-run marginal costs associated with additions to capacity on a system appear to be fairly constant over time.[63] Economic efficiency requires that the tolls be as close as possible to marginal cost. The front-end loaded pattern associated

62 For example, in the case of Group 1 oil and products pipelines, the NEB currently employs light-handed, COS regulation with a banded rate of return.

63 This concerns full-cycle incremental costs. The shorter run marginal costs will usually be low if the expansion is via compression and they will usually be higher if it is via looping. See the evidence presented by TCPL in NEB Hearing GH-5-89.

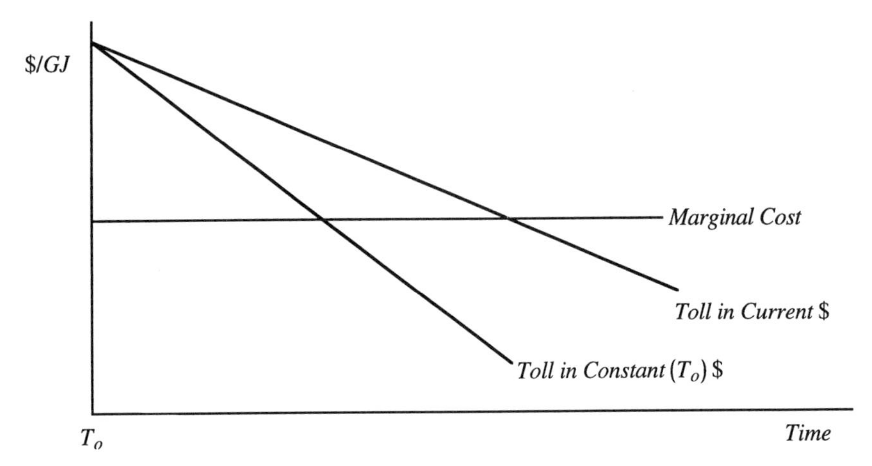

FIG. 4.1 TEMPORAL TOLL PATTERN UNDER STATIC CONDITIONS USING COS REGULATION

with COS tolling on a stand alone basis results in significant inefficiency and a redistribution of benefits from initial customers to later customers. That is, as shown in Figure 4.1, the tolls will be well above marginal cost in the early years and well below marginal cost in the later years. However, these negative consequences will be substantially reduced if the system expands over time and rolled-in methodology is employed.[64] As illustrated in Figure 4.2, the effect is to level the temporal pattern of tolls. The additions to the rate base cause tolls to rise such that the decline that would otherwise be observed because of the front-end load is largely offset.

A point which is particularly important in the context of incentive schemes such as price caps is that the temporal toll patterns will be to a large extent determined by the rate of capacity expansion. In cases where there are few significant expansions, the real tolls will tend to drift downward due to a depreciation and a declining rate base. On the other hand, if there were frequent expansions, a more constant real toll pattern would be expected. Consequently, unlike the situation envisaged under many price cap schemes where real tolls or prices are reduced over time through productivity change, here the temporal toll pattern tends to be driven, not by productivity growth or technological

64 That is, the costs associated with system expansions are rolled into the existing rate base so that there remains one toll for each type of service. This is unlike vintage or incremental tolling, where each expansion is tolled separately with the result that there are different tolls for the same service based on differences of the vintage of the equipment used to provide the service.

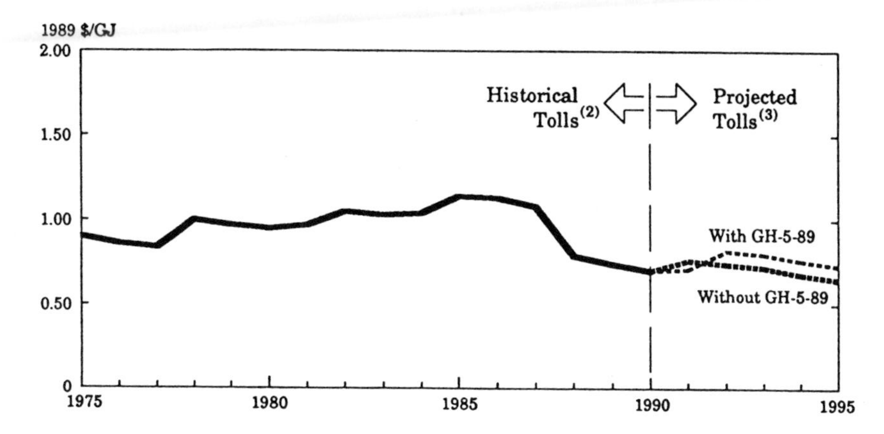

Source: Page 12 of NEB (1990a).

FIG. 4.2 HISTORICAL AND PROJECTED TCPL TOLLS TO EASTERN ZONE

change, but by simply the rate at which additions to capacity are made. A price cap scheme which requires declining real tolls over time may, therefore, be quite inconsistent with requirements for significant capacity expansions.

Alternative Depreciation Methodology. It may be theoretically possible to lower real tolls even in an expansion environment by changing the depreciation pattern and, consequently, the pattern of capital recovery. For example, there could be a shift to a more level depreciation pattern leading to 'levelized' tolls or, in the extreme opposite case to front-end loading, there could be a shift to a more delayed recovery of capital which leads to 'back-end loaded' tolls.[65]

There are several important issues with regard to these alternative depreciation and toll patterns. The first one concerns which particular depreciation pattern is optimal from a societal viewpoint. For instance, one does not observe front-end loading in many unregulated prices and this may suggest that front-end loading is not optimal. Two approaches have been used to address this issue. The one employed by Baumol and Bradford (1970) determines the optimal depreciation pattern based on the optimal path for prices. The more recent approach associated with

65 The use by the InterCoastal Pipeline Project of the Reverse Sum of the Years Depreciation (RSYD) represents a recent example of a shift to a different depreciation methodology to increase the competitiveness of tolls in the near term. See the application by InterCoastal Pipe Line Inc. to the NEB, Hearing Order GH-4-93.

Burness and Patrick (1992) calculates the optimal recovery of capital costs by a profit-maximizing firm operating under rate-of-return regulation and by an economic welfare maximizing regulator. It will suffice to note that these studies show that there are conditions, such as rapid technological change of the type experienced in the telecommunications industry, where depreciation patterns that produce front-end loading are optimal. In cases like these it is necessary to recover the capital invested earlier rather than later because cost-reducing technological change will make a delayed recovery of capital very difficult to achieve. However, these studies also suggest that, under more commonly observed conditions, depreciation patterns that generate prices that are more level over time, or even somewhat back-end loaded prices, are closer to optimal.

The second issue has to do with the effect of changes in depreciation or capital recovery patterns on the cost of capital. If, for example, the shift to a methodology which delays the recovery of capital significantly increases the perceived risk faced by investors, the result may not be a reduction in tolls. The increase in the cost of equity capital and / or the necessity of moving to lower debt / equity ratios could offset the toll reductions achieved by reduced front-end loading.

Incremental versus Rolled-In Tolling. Under rolled-in tolling, all expansion costs are rolled into a single, existing rate base. As such, all customers using a given service pay the same toll regardless of the length of time each has been a customer on the system.

'Incremental' is typically defined in the economics literature as the discrete version of marginal. For example, 'incremental cost' is the discrete version of 'marginal cost.' Following this definition, incremental tolling would be interpreted as a situation where tolls or prices are set equal to incremental or marginal cost.

Unfortunately, the term 'incremental tolling' has come to mean something quite different in an applied, regulatory setting involving the regulation of pipelines. As used there, incremental tolling is really a form of vintage tolling where there is a separate rate base and corresponding tolls for each vintage of facilities constructed after some given year. That is, the toll charged a particular customer depends on when that customer joined the system. If the customer was deemed to be in the 'old vintage' category, the toll would be based on the facilities with the lowest embedded cost; that is, these users would in essence be assigned the facilities against which the largest accumulated depreciation expense had been charged. At the other extreme, the newest customers would be charged a 'new vintage' toll based on the undepreciated cost of the latest facility additions. It is perhaps not surprising that in most cases, incremental or vintage tolling schemes have been proposed by existing (or 'old vintage') shippers wanting protection from increasing tolls due to capacity expansions. It can be observed that these vintaging approaches lead to many different tolls for the same service and that, since there will only be one marginal cost for that service, they cannot all be equal to marginal cost. In contrast, with true incremental or marginal cost pricing as used in the economics literature, all customers face the same price for a given service and this price is set to approximate the marginal cost of providing that service. This comparison makes it clear that the vintage pricing typically proposed under the heading of 'incremental tolling' is really a form of price discrimination based on the vintage of the customer demands for service.

Most incremental or vintage schemes involve drawing a line or boundary where everyone within it will enjoy the declining real tolls associated with front-end loading, while those outside the boundary will face much higher tolls for the same service, but these too will decline over time assuming all future expansions are also vintage tolled. It can be observed that the appeal of incremental or vintage tolling is directly related to the front-end loading in toll patterns as outlined above. If, for example, levelized tolling was employed, it would be difficult to mount a case for incremental tolling. That is, there would typically be only minor differences in levels between rolled-in tolls and the various vintage tolls.

An evaluation of incremental versus rolled-in tolling in the context of natural gas transmission can be found in the NEB's November 1990 decision (NEB 1990a) and in Mansell (1994). The following is a brief summary of the main arguments in the context of an expansion on an integrated system.

Temporal Toll and Cost Patterns. As noted earlier, rolling-in expansions can assist in producing a more level temporal pattern for real tolls. Depending on the rate of expansion, the deviations of the rolled-in toll from marginal cost can be relatively small and constant over time.[66] Tolling any significant expansions on an incremental or vintage basis is likely to result in large differences in tolls for identical service. Further, the new vintage toll for the latest expansion will initially be much higher

---

66 Here and in what follows in this section, references to marginal cost mean references to the marginal cost of expansion for an existing system. See text at footnote 62 for further explanation.

than marginal economic cost and the old vintage toll will typically be below marginal cost and fall further below this cost over time.

Fairness and Equity. An important argument typically employed by advocates of incremental or vintage tolling is that it is the new customers requesting service who cause the need for the new facilities and, therefore, it is they who should be responsible for any increased tolls associated with the new facilities. However, as emphasized in NEB (1990a), the validity of such arguments is questionable. First, in many cases it can be demonstrated that all system users (including old users who maintain or renew their demands) 'cause' the need for additional facilities and, therefore, all are responsible for the associated increased costs. "With regard to the debate as to who caused the need for new facilities, the Board is persuaded by the argument that it is the aggregate demand of all shippers that gives rise to the need for additional pipeline capacity." (NEB 1990a, 13)

Second, in the case where a utility simply provides a service, such as gas transmission, there are no acquired rights. "... [T]he payment of tolls confers no future benefit on tollpayers beyond the provision of service. In other words, previous tollpayers have no acquired rights. Therefore, they cannot expect to be exempted from a toll increase simply because they paid tolls in the past." (NEB 1990a, 12)

In contrast, rolled-in tolling does not assume that old vintage shippers have rights to the old facilities with their lower embedded cost and lower value because of depreciation. Consequently, rolled-in tolling does not imply 'rights' to the low tolls that would be associated with the old, depreciated facilities. Further, this approach is consistent with the notion that all system users 'cause' the need for new facilities.

The argument that vintage tolling is required to prevent an inequitable redistribution of income has questionable validity in many, if not most, cases. That is, it is suggested by the proponents of vintaging that existing shippers deserve all of the benefits associated with the continually declining tolls for old vintage shippers. The implicit assumption is that all those who would benefit from these declining tolls were in fact those who paid the front-end load. Factors such as the previous and ongoing turnover in the ownership of capacity rights due to restructuring, open access and capacity release would generally make it extremely difficult to establish which shippers were the historic customers and which shippers had paid the original front-end load and, therefore, deserve the lower old vintage tolls. The alternative of rolled-in tolling involves a sharing among old and new users of a 'subsidy' paid by those in the initial years who carried the burden of the front-end load in the toll structure.

It is also worth noting that the arguments by proponents of incremental tolling tend to be asymmetric. For example, they typically argue that vintage tolling should be applied in the case of a 'looping phase' expansion (which tends to be more expensive than an expansion via added compression and which is likely to raise unit costs and tolls) but not in the case of a 'compression phase' expansion (which can often result in a decline in unit costs and tolls). In the latter case, a roll-in is usually suggested so as to lower unit tolls for existing (and new) shippers.

Rate Stability. Rolled-in tolling can result in a significant short-run increase in tolls when there is a large expansion to a highly depreciated system. However, in determining whether the toll increases associated with even large expansions constitutes rate shock, it is necessary to evaluate these toll increases in the context of longer-term trends. For example, in the case of the GH-5-89 facilities, the NEB noted that the first year impact of the roll-in would be to increase the Eastern Zone toll by 33 percent, but even with this increase the toll would be lower in real terms than it was two decades earlier. (NEB 1990a, 11). To the extent there is a problem of rate shock in particular instances, it may be preferable to deal with it through some form of partial, short-run levelization than to attempt to resolve it by abandoning rolled-in tolls and implementing vintaging. The negative effects of vintaging in terms of other criteria might far outweigh any positive effects of protecting a subset of system users from toll increases. Further, vintaging results in greater rate shock for new markets and these may be the most sensitive to significant price shifts.

Proper Price Signals and Economic Efficiency. Economic efficiency requires that the price or toll for the capacity be as close as possible to marginal economic cost. Because of front-end loading, the initial tolls for new users under incremental tolling will typically far exceed long-run marginal cost, while the tolls for old users will fall increasingly below marginal cost. Since, the renewals of demand for the service by the old users are also a cause of the need to expand, this produces inefficiency because it signals to them that marginal cost is much lower than it actually is. Further, there is the problem of conflicting signals. Some (the new users) receive the signal that marginal cost is very high while others (the old users) receive the signal that marginal cost is very low. As the number of expansions increases, the number of different tolls for the same service expands correspondingly, as do the number of signal errors and cases of allocative inefficiency.

Another flaw is that this methodology assumes that the signal received by the market will, at the margin, be the new vintage toll.

However, in many cases the shipper would be an existing buyer such as a Local Distribution Company (LDC) that would simply roll the old and new vintage rates together. The ultimate purchasers (the customers of the LDC) would, in any case, see this rolled-in toll and not the new vintage toll. Other customers, such as cogeneration plants, may not have the ability to roll-in or average down the new vintage toll. This, in turn, would introduce further unwarranted and unintended discrimination and inefficiency.

Rolled-in tolling has one major advantage in that it embodies uniform pricing so that all customers receive the same price signal. Whether this one signal is close to marginal cost may depend to a significant degree on whether the system has been static or expanding. If it has been expanding over time, the rolled-in toll may be reasonably close to marginal cost and closer to marginal cost than tolls under incremental methodology. In this case, all customers will be receiving close to the correct signal in terms of the requirements for economic efficiency. If, however, the system has not been expanding so that the rolled-in toll is much below marginal cost, it will be important to take this signal error into account in determining the best allocation of existing capacity and how much new capacity can be economically justified.

In cases where the rolled-in toll deviates markedly from marginal cost, it can be shown that efficiency gains are possible by employing Ramsey pricing. This involves discriminatory pricing such that a price close to marginal cost is charged in the most price sensitive markets while offsetting prices which deviate markedly from marginal cost are charged in the least price sensitive markets. The general idea is to get the market signal as close as possible to the right level for those markets where it matters most, while worrying less about it in the more price-inelastic markets.

Advocates of incremental or vintage tolling often point to these Ramsey pricing arguments for support. However, it is not clear that the particular price-level and elasticity combinations under vintage tolling would generally correspond to that required for Ramsey pricing. For example, it may be that many existing shippers such as those with only short-term contracts or those associated with industrial markets (and who would benefit from low tolls under vintage pricing) constitute a more price-sensitive market component than new shippers (who would face the highest tolls under vintage pricing). This corresponds to the TCPL expansion dealt with in NEB Hearing GH-5-89 (NEB 1990a). With vintage pricing, the tolls for the former group would be below marginal cost and fall even further below this benchmark over time. On the other hand, the tolls for the latter group would be well above marginal cost initially, before declining over time. But with Ramsey pricing the tolls for the former group should be higher (closer to marginal cost) than under the vintage approach and thus the tolls for the second group are likely to be lower than with vintage tolling.

However, it is more likely the case that new markets are more price-sensitive than existing markets. Ramsey pricing would then imply that prices in the new market should be less than those for existing markets.[67] Advocates of vintage or incremental tolling argue that the elasticity in new markets is greater simply because the new markets have yet to make a commitment in the form of investment in the infrastructure or other equipment related to a particular alternative. That is, *ex ante* substitutability, and hence price sensitivity, is greater than *ex post* substitutability. However, using this reasoning in the context of Ramsey pricing would mean that as soon as the new market is connected, price elasticity will be decreased and, therefore, the justification for the discrimination based on vintage disappears. In any event the price sensitivity in new markets is also likely to be greater because of the presence of competitive substitutes such as the existing energy sources.

Revenue Sufficiency and Stability. Both approaches can satisfy the revenue sufficiency criterion. Any differences are likely only to be significant in terms of the relative stability and predictability of the revenue stream. Rolled-in tolls are likely preferred in terms of this criterion since they involve an important element of diversification.

Consistency With Other Policies and Regulation. Incremental or vintage tolling results in new vintage tolls which are above marginal cost and above the toll under rolled-in methodology. This generally results in artificial and significant hurdles in expanding transportation to new markets. If the new customers were served by an aggregator such as an LDC, the hurdle would be smaller since the aggregator would have the ability to roll together or average the various vintage tolls.

Depending on the nature of the expansion, the older vintage tolls could be significantly lower than a rolled-in toll. However, this may not result in sufficient gains in gas consumption in markets served under old vintage tolls to offset the lost opportunities in new markets subject to the much higher new vintage tolls. For example, the new markets would typically be more price-sensitive because no commitments have yet been made in the form of the equipment required to use a particular type of fuel. Put differently, *ex ante* fuel substitutability is greater than *ex post* substitutability.

---

67 Assuming that the marginal cost in each market is the same.

Practicality, Administrative Simplicity and General Acceptance. Rolled-in tolling generally meets these criteria. This methodology, and its application, has a long history, there is no need to discriminate among system users receiving the same service, and the approach is easily understood.

While the information and computer technologies available may make it possible to design and implement the more complex pricing systems associated with incremental or vintage tolling, there can be other problems. These can include the difficulties associated with ad hoc rules to separate old and new vintage users since criteria such as 'historic tollpayer' or 'those who suffered the incidence of the front-end load in the initial years' cannot be defined or applied accurately; the necessarily ad hoc rules to allocate maintenance, repair and operating costs among the many vintage rate bases; continually having to defend wide variations in tolls for identical service; and defining the rights of existing users and the prospective rights of new users without granting acquired rights. As with any approach, the regulatory and administrative costs would increase with the growth in controversy. However, it is useful to note that in GH-5-89, the NEB did not reject any of the incremental or vintage alternatives on the basis of impracticality or lack of simplicity (NEB 1990a, 13).

Secondary Capacity Implications. A well-functioning, competitive secondary market for gas transmission capacity rights can play an important role in the efficient allocation of capacity. Further, such a market can reduce the signal errors and inefficiencies associated with both tolling methodologies. For example, if there is a sizable, active, and competitive secondary market, the effective tolls established in that market may be closer to marginal cost than the rolled-in or incremental tolls established in the primary market. The tolls established in the secondary market (which could be above or below the tolls established in the primary market) may also provide better signals for efficient capacity expansion. For example, if the effective tolls set in the secondary market were consistently above the marginal cost associated with capacity additions, this would provide a clear signal that an expansion was warranted on efficiency grounds.

However, there is also the likelihood that, relative to the case of rolled-in tolling, vintage tolling can result in greater complexities and problems with respect to the secondary market. For example, consider the situation where there are no binding caps so that secondary capacity could trade at a price up to the highest incremental toll.[68] In this situation, the observed price in secondary markets would typically be below this limit given the existence of considerable capacity tolled at much lower, old vintage tolls. This would mean additional risk to those shippers required to underpin new facilities by signing long-term transportation contracts. Any market changes that entail capacity release would usually involve large financial losses for the new shippers. Further, the competitive advantage to those able to secure capacity on the secondary market at a price less than the new vintage toll would increase the probability that the new vintage shippers would lose markets and have to resell capacity at a significant loss. These factors would make it more difficult to finance significant expansions to capacity since shippers would be more reluctant to sign the long-term contracts required to underpin the investment.

Tolling and Optimal Investment. To the extent that the new vintage tolls are significantly above marginal cost (because of front-end loading), incremental or vintage tolling will generally have the effect of discouraging the construction of facilities that do meet the test of economic efficiency. Further, it exacerbates the problem that, if the existing shippers were given a more appropriate price signal, some of the low-value use of existing capacity would decrease and this would minimize the need to construct new facilities.

In the case of rolled-in tolling, there are also possibilities of deviations from optimal investment. For example, the existing toll for a highly depreciated system may be considerably below marginal economic cost. The ability to roll-in new facilities could produce a tendency to over-build capacity. However, there may be two checks against this. First, one of the roles of the regulator is to ensure that proposed facilities meet necessity and efficiency criteria. That is, there usually exist implicit or explicit regulatory mechanisms to control any overbuilding tendencies. Second, as illustrated in the GH-5-89 case involving expansion of the TCPL system, the rolled-in toll for an expanding system may in fact be reasonably close to marginal cost. In such cases, the inherent tendency to overbuild would be significantly reduced.

In summary, for many of the reasons outlined above, Canadian regulators have generally favoured rolled-in tolling over incremental or vintage tolling for expansions on integrated systems. It should be emphasized, however, that this conclusion does not necessarily apply to other situations such as those involving the construction of laterals.

---

68 Note that the NEB has recently declared that there should be no price caps for capacity trading on the secondary market.

For example, it is not uncommon to find incremental tolling for downstream laterals that are dedicated to providing service for one user or an identifiable group of users. Similarly, elements of incremental tolling (such as contributions in aid of construction or surcharges) are frequently applied setting tolls for upstream laterals if it is unlikely that the full incremental cost of such facilities will be recovered over their expected life under rolled-in tolling.

While regulatory decisions in Canada have generally been in favour of rolled-in tolling for major system expansions, there has been less consistency in the U.S. For example, the decision by FERC to employ incremental tolling for the GH-5-89 expansion on the Great Lakes Gas Transmission (GLGT) system threw into question the previous commitment to rolled-in tolling in the U.S. This and other decisions have been under appeal and the recent Court decision concerning incremental tolling on the GLGT system required the FERC to reconsider the use of vintage tolling.[69]

In response to this ruling, the FERC set up a hearing to receive input.[70] Subsequently, it issued a statement of policy aimed at achieving the following principal goals: "... to provide the industry with as much upfront assurance as is possible with respect to rate design to be used for an expansion project, while, at the same time, to provide for a flexible assessment of all the relevant factors of a specific project... [and] specific attention should be paid to efforts to minimize significant rate shocks on existing shippers that may be produced from rolling-in the costs of expensive projects." (FERC 1995, 4)

The policy adopted by the FERC has two main features. "First, the Commission will make a determination of an appropriate rate design in a pipeline's certificate proceeding. Second, when the pipeline seeks rolled-in pricing, the Commission will base its pricing decision on an evaluation of the system-wide benefits of the project and the rate impact on existing customers."(FERC 1995, 4) "[To reduce uncertainty] the Commission will establish a presumption in favour of rolled-in rates when the rate increase to existing customers from rolling-in the new facilities is 5 percent or less and the pipeline makes a showing of system benefits."(FERC 1995, 6) These benefits could be of the operational

---

69 See TransCanada Pipelines Ltd. v. F.E.R.C., 24F. 3d 305 D.C. Cir. (1994). Briefly, the Court concluded that the use of incremental tolls violated accepted criteria concerning 'just and reasonable' tolls and 'no undue discrimination.' It also concluded that the 'commensurate benefits' test used to justify vintage tolling is not a traditional or generally accepted approach and that inadequate justification had been provided for a change from rolled-in tolling.

70 See FERC (1994).

variety (such as increases in system or operational reliability, increasing shippers' access to new supplies or markets, or providing greater flexibility to meet imbalances) or of the financial variety (such as enabling the pipeline to serve new demands to replace demands being lost to the system, reducing customer costs, or producing economies of scale in the form of reduced future expansion costs). If these benefits are shown and the rate impact is 5 percent or less, "... customers opposing rolled-in rates will have the burden of showing that the benefits are so insignificant that rolled-in rates are not justified. When the rate impacts exceed the 5 percent threshold, the presumption will not apply and the pipeline will need to show that the benefits are proportionate to the rate impact. Pipelines should not break projects into small segments solely to qualify for the 5 percent test for each project."(FERC 1995, 7)

Other important aspects of this policy are: there is a presumption in favour of incremental or vintage tolls for downstream laterals that are for the benefit of one or a few customers; in the case of upstream supply area laterals, which often provide greater access to supplies for all customers of the system, tolling would be determined according to the principles noted earlier for mainline capacity; and considerable attention should be paid to mitigating general rate increases as a result of roll-ins (for example, by phasing-in the roll-in) and to ensuring that any new capacity is put in place only when it is clear that there is insufficient release capacity to meet the new demands.

4.4 ASSESSMENT OF COS REGULATION

Allocative or Pricing Efficiency. Allocative efficiency is achieved when the optimal amount of capacity is provided. The price level under COS regulation is set by dividing the revenue requirement by quantity. To the extent that the revenue requirement is determined on the basis of historical cost concepts (especially the imputation for depreciation) and not opportunity costs, the resulting prices will not be optimal since they will typically deviate from the marginal or incremental cost of capacity. In a multiproduct firm, the use of fully distributed cost pricing and historic accounting principles does not generate Ramsey (second-best) prices.[71]

Also, these prices may not provide appropriate signals to shippers regarding the cost of providing additional capacity. These inefficiencies are partially off-set to the extent that two-part tariffs (fixed/variable tolls) are employed optimally. Further, there may be cases where capacity expansions, combined with rolled-in tolling methodology,

71 See Section 3.3.

results in prices that are fairly close to marginal cost. For example, in NEB GH-5-89 it was demonstrated that the tolls for TCPL have been reasonably close to the long-run marginal cost of expansion. The conclusion is that pricing inefficiency is likely to be much more of a problem under COS regulation if the system is static or if it is expanding but incremental or vintage tolling is employed. Finally, as noted in Section 4.3, there is the potential even in a static system to improve pricing efficiency under COS regulation through altering the depreciation patterns employed in determining costs and setting tolls.

Efficient Rationing. Efficient rationing requires that existing capacity be used optimally. This means that the pricing structure and conditions of service be flexible enough to allocate capacity optimally, both when there is insufficient capacity and when there is surplus capacity. When there is surplus capacity, an optimal allocation exists if no capacity goes unused provided a shipper is willing to pay an amount at least equal to variable cost. An optimal distribution under conditions of insufficient capacity occurs when the existing capacity is allocated to those who place the highest value on it.

Traditional COS regulation typically handles the case of excess capacity much better than the case of insufficient capacity. Most pipelines offer some form of interruptible service. This provides some downward price flexibility when there is excess capacity. However, when there is insufficient capacity, the curtailment procedures used generally do not ration on the basis of value or willingness to pay.

Product Selection. Optimal product selection implies a close match between available services and the preferences of customers. To the extent that all firm service (implicitly) has the same level of reliability (probability of service), traditional COS regulation does not allow shippers to trade off price for reliability or quality. The same is typically true for interruptible service when there is only one or two types of interruptible service.

On the other hand, the overall level of reliability under COS regulation is very high. Firms subject to COS regulation do not have an incentive to reduce the quality of service. Reductions in the quality of service will not increase profits since any cost savings will be passed on to consumers. However, it will be in the interests of management and regulators to avoid the complaints and embarrassment associated with poor quality of service. Finally, if the quality of service is a function of operating costs or capital, a positive side effect of any cost inefficiency may be a high quality of service.

Cost Efficiency. Critics of COS regulation contend that it provides perverse incentives for cost minimization. They argue that such regulation provides firms with no incentive to control managerial inefficiency and it provides them with an incentive to over-invest in capital.

Managerial inefficiency typically takes the form of excessive operating costs. It is argued by critics that the lack of incentive for a firm subject to COS regulation to reduce operating costs follows from the link between prices and costs. If the costs of the firm are excessive under COS regulation, the prices charged by the firm will also be excessive.

Moreover, it is argued that the firm does not have strong incentives to reduce its costs. It is suggested that since reductions in costs lead to lower tolls, the firm is not rewarded for reducing costs. The increase in operating costs due to managerial inefficiency can take many forms. Some examples are excessive layers of management, failure to negotiate 'hard' with factors of production, underutilized inputs, and excessive perquisites.

Averch and Johnson (1962) presented a model of ROR regulation where the allowed rate of return exceeds the rental rate or cost of capital. They used it to show that under this form of regulation firms have an incentive to increase the size of the rate base. A larger rate base increases profits since an extra dollar of total capital costs generates increases in revenues of more than a dollar. The result is that the firm will utilize its inputs inefficiently; that is, it will bias its input ratios in favour of capital expenditures. For instance, if the firm only uses capital and labour, the capital / labour ratio will be too high. This means that the firm could produce the same level of output at a lower cost if its input ratio was closer to the efficient level.

In the context of pipelines, this inefficiency could show up in two forms. The first is that pipelines may have an incentive to overbuild their systems. As a consequence, capacity expansions may be premature and excessive. The second is that pipelines may have insufficient incentives to control the costs of capacity additions if any resulting cost overruns are simply included in the rate base.

Joskow (1974) has stressed that the Averch-Johnson (AJ) model assumes that regulators delegate pricing responsibilities to firms and that the regulator continually enforces the constraint that the actual rate of return equal the allowed rate of return. Joskow points out, on the other hand, that in reality pricing is usually not delegated to firms and that the regulatory constraint on the allowed rate of return is not continually monitored. This suggests that the implications of the AJ model are best viewed as possibilities in the more common case of COS (rather than ROR) regulation.

There are two features of COS regulation which, in practice, do give a regulated firm incentives to minimize costs. First, Joskow and others have argued forcefully that regulatory lag—the time between rate hearings and changes in costs—provides firms with an incentive to cost minimize. Any reduction in costs between rate hearings increases profits that the firm may be allowed to keep under certain forms of COS regulation, at least until the next rate hearing when prices are adjusted again to reflect costs. This may be particularly effective if, as Joskow finds, regulators and customers are less likely to initiate rate hearings. This means that COS regulation is only cost-plus regulation in the long run: in the short-run, prices are fixed. Nevertheless, it should be noted that the strength of this incentive will be diminished if the firm must appear before the regulator on an annual basis to have tolls adjusted. In such a short time-frame, activities to achieve efficiencies may be limited to those with one year pay-backs.

The second feature is that regulators typically have the power to disallow imprudent investments and costs. An integral part of COS regulation is the prudence review under which all current and capital costs incurred by the firm are subject to review. Lyon (1991) shows that hindsight reviews do have desirable efficiency properties in a model of COS regulation similar to that used by Joskow. He shows that prudence reviews reduce the tendency of firms subject to COS regulation to invest in large risky projects and, consequently, the level of investment moves closer to that dictated by cost considerations.

The argument that COS regulation results in overbuilding also neglects certain features of the implementation of this type of regulation as practiced by regulators of pipelines in Canada. The need for significant expansions to capacity must usually be demonstrated by long-term contracts for the extra capacity which have been signed by shippers. This provides considerable market discipline given the demand charge obligations embodied in such contracts.[72] This market discipline is further enhanced when appropriate consideration is given to the reality that pipelines compete for expansions and, consequently, have incen-

72 It should be recognized, however, that this discipline may not be efficient if the toll for the new shippers deviates significantly from marginal cost. For example, if it was much below the incremental cost of capacity, this signal would lead to more requests for new capacity than would be optimal. If this toll was above marginal cost, the requests for new capacity would be less than the optimal level.

tives to create and maintain a reputation with shippers for efficiency. Also, in most cases the regulator imposes additional tests for determining whether extra capacity is warranted.[73]

Gilbert and Newberry (1994) argue that while considerable attention has been devoted to the incentive for over-investment under ROR, its "... advantage in constraining opportunistic behaviour appears to have been overlooked." (Gilbert and Newberry 1994, 541) They focus on the sunk nature of investment by utilities and the possibility that the regulator could expropriate the capital investment of the firm. The regulator can hold up the firm after it has invested by allowing it a less-than-market return on its capital. They argue that the recent data on capital investment by utilities in the United States suggests that, after the tightening of the regulatory constraint in the 1970s and 1980s, firms have responded by under-investing in capital in order to minimize the risk of being held up by regulators who favour consumer interests. They demonstrate that following rate-of-return regulation on investments which are 'used and useful' is a superior regulatory regime because it makes it less likely that the regulator will attempt to hold up the firm. This means that a firm subject to this type of regulation will invest efficiently and not under-invest in capital.

Incentives for Investment in Cost Reductions. In a dynamic context it is alleged that firms under COS regulation do not have very strong incentives to adopt new technologies that will reduce costs since any reduction in costs is entirely passed on to consumers in the form of lower rates.

However, there are two mitigating features. The first is that, to the extent that cost-reducing innovations are embedded in capital inputs, the firm (due to any existing Averch-Johnson effects) may have ample incentive to invest in cost reductions. Thus, to a certain extent, there may be positive long-run consequences associated with AJ effects. It is expected that, in the case of pipelines, technological change will often be embedded in capital inputs. Secondly, to the extent this is true, the rate of technological change is more likely to be driven by the ability of the regulator to commit not to expropriate the capital investment of the utility.[74]

73 For example, see the extensive list of considerations used by the NEB. These are outlined in any of the recent decisions concerning capacity expansions for Group 1 gas pipelines.

74 This result follows from the discussion in the preceding section regarding under-investment.

Lyon (1995) demonstrates that the rate of innovation in the U.S. electric utility industry has slowed remarkably since the early 1970s. His explanation for this decline is that it is the response to expropriation by regulators of sunk investments made by U.S. electrical utilities through prudence reviews. He shows that the relatively recent move by regulators to prudence reviews where investments are disallowed based on 'cost-effectiveness' has reduced the incentives for utilities to invest in risky new technologies. Cost effectiveness means that investments are admitted into the rate base not on the basis of whether they are used and useful, but rather on the basis of whether or not the expenditures were 'prudently incurred.' His analysis supports the conclusions of Gilbert and Newberry (1994): if a regulatory regime does not protect the firm against capital expropriation or hold-up, firms will under-invest in capital.

Fairness and Equity. This criterion has two main aspects. The first is that tolls must be just and reasonable. In practice this usually means that tolls are cost-based, that there is no undue discrimination (including price discrimination), and that rates are reasonably stable (that is, there is an absence of 'rate shock'). The second is that cost reductions benefit customers.

Traditional COS regulation rates quite highly in terms of both elements of fairness. In general, it provides regulators with the flexibility to arrive at a rate structure which meets fairness criteria. Tolls are usually based on fully distributed costs, price discrimination is typically not allowed, and rolled-in tolling (perhaps combined with some levelization or phasing-in of the increased costs associated with expansions) can provide a high degree of rate stability and predictability. Also, by its very nature, COS regulation passes reductions in costs to customers through lower prices.

Economic Viability. This also has two components. The first concerns the viability of the firm. Does the regulatory regime put the long-term viability of the regulated firm at risk? The second is that the regulatory regime should be sustainable or stable over lengthy periods of time. The operation of the regime should not result in outcomes so extreme that dramatic regulatory shifts become unavoidable.

On both of these counts COS regulation scores well. As noted earlier, COS regulation has generally been applied in the context of a 'regulatory compact' which protects the recovery of approved investments by the firm. In most cases, the shareholders face little risk. Also, as demonstrated by the long-term use and acceptance of COS regulation for natural monopolies, it is sustainable. In most cases where it has been abandoned, regulation was either not economically justified in the first place or it became unnecessary because of technological or other fundamental changes.

Regulatory Burden. The regulatory burden of COS regulation is fairly significant and often a lightning rod for discontent. Rate proceedings are generally an adversarial process since shippers and the pipeline can affect the overall level of tolls if they can influence the determination of allowed costs, the size of the rate base, and the allowed rate of return. Different classes of shippers have incentives to attempt to alter the rate structure in ways that serve their particular interests. Spulber (1989) has suggested that rate hearings are more correctly viewed as an institution which allows customers and firms to bargain over the allocation of economic rents. In such a framework, both groups have an incentive to engage in rent seeking; that is, to incur expenditures to try to influence the decision of the regulator. These types of costs significantly increase the cost of COS regulation above and beyond the direct administrative costs of the regulatory institutions.

Spulber contends that the COS hearing is an attempt to minimize transaction costs between firms and their customers. It provides an institutional framework within which bargaining between firms and customers can occur. Such hearings can be interpreted as a set of rules for negotiation and dispute resolution, with the regulatory board adopting the role of arbitrator in the event of an impasse. His contention is that the private contacting costs associated with governing the commercial relationship between the firm and all of its customers on an individual basis would exceed the costs of regulation.[75]

In any event, these costs may be a necessary component of a system which is perceived to follow a process which is fair and whose outcomes are also perceived to be legitimate. Openness and public participation by all interested parties in decisions which materially affect them would appear to be a requirement of any democratic regulatory structure.

The frequency of hearings in the Canadian pipeline industry in the 1980s and the early 1990s reflects the many capacity expansions, especially since the deregulation of gas markets and prices. Additions to the rate base require a rate determination to adjust tolls. When capacity additions are lumpy, there may not be any other alternative but to have a rate determination each time capacity expansions are

75 This is related to the discussion of regulation as a complement to private contracting as a rationale for regulation in Section 3.2.

included in the rate base. The costs associated with rate proceedings are greater if there are large facility expansions. In the absence of large facility expansions, rate hearings are typically routine and less costly.

In any case, as argued below, there are various streamlining options that can significantly reduce the regulatory burden and also increase the acceptance and effectiveness of COS regulation.

Implementation. This encompasses the notions of practicality, administrative simplicity and general acceptance. COS regulation is fairly easy to implement since it mostly uses readily accessible and verifiable accounting data. Of course the determination of which costs are to be included in determining the revenue requirement will usually be a contentious issue. The fact that it has been used for many years means that most participants have a good understanding of the process.

General Assessment of Cost of Service with Streamlining. As outlined in Section 4.2, there are a number of institutional modifications that fall under the heading of streamlining. These include significant use of task forces and committees, the employment of negotiated settlements, exemptions from public hearings for routine investments or later investments which are part of a multi-year expansion program, and the use of formulas or generic hearings to set allowed rates of return. As noted in Section 4.2 (under *Streamlining*), the NEB has recently initiated a series of streamlining initiatives, including the development of a generic, formula approach for setting the return on equity for a number of pipelines; the encouragement of the use of negotiated settlements and the establishment of guidelines for such settlements; and the expedited treatment of certain routine and multi-year expenditures.[76]

There is the potential that streamlining can significantly reduce the level of regulatory burden under COS regulation and increase overall regulatory effectiveness. The use of less formal mechanisms such as task forces and committees may allow greater participation in the regulatory process. Initiatives which result in increased time between formal hearings can also increase the incentives for efficiency associated with regulatory lag. Further, since a considerable component of a typical hearing is taken up by the determination of the allowed return on equity, measures like the generic, formula approach instituted by the NEB have the potential to significantly reduce the time and cost associated COS regulation.

---

76 On the last, see NEB, File No. 3600-A001-15 (July 6, 1994), Re: Section 58 Streamlining Initiative.

However, unless the streamlining initiatives are designed and implemented carefully, there can also be some disadvantages. These include possible reductions in the degree of regulatory oversight and the perception by some stakeholders excluded from the process that the procedures are less fair and the outcomes are less equitable. The degree of regulatory oversight might be reduced if the regulator felt compelled to honour all negotiated settlements. Problems with negotiated settlements may arise because of asymmetries of information, expertise, and bargaining power between pipelines and the shippers. Moreover, differences in the interests between different types of shippers (large vs small, new vs old, north vs south) may make a unified industry perspective hard to maintain. There are very difficult interpretation problems when the negotiated settlement does not enjoy universal support from shippers. Finally, there is the problem of strategic behaviour on the part of both the firm and the shippers. Either one has an incentive to arrive at a negotiated settlement and then later introduce before the regulatory board new issues not considered by the settlement. Depending on the nature and resolution of those issues, support for the original settlement may evaporate.

5 *Incentive Alternatives*

Incentive regulation is aimed at improving the performance of the regulated firm through the use of rewards (and penalties) beyond those incorporated in traditional COS regulation. For example, it might involve allowing the firm to earn a higher rate of return if its score in terms of some performance measure improves. The main attempt is to break the link between costs and prices so that the firm has greater incentives to reduce costs. That is, if prices are not simply determined by adding up the costs incurred, the regulated firm can achieve higher profits by cutting costs since prices will not automatically be reduced by the cost savings.

Section 5.1 highlights some general considerations in designing and implementing incentive regulation. The following section (Section 5.2) outlines incentive schemes which involve regulating the price level rather than the rate of return or profits. These schemes include Price Caps (PCs), Automatic Rate Adjustment Mechanisms (ARAMS), Sliding Scale Plans (SSPs), and Yardstick Competition (YC).

Section 5.3 sets out the main incentive alternatives that involve some version of profit regulation. That is, under these schemes, the profits of the regulated firm will vary depending on the behaviour of the firm but the prices for the products or services provided are still, at least in part, determined by cost levels. These mechanisms are Profit Sharing (PS), Banded Rates of Return (BROR), Benchmarking (BM), and Capital Cost Incentives (CCIs). Other schemes aimed at improving efficiency, including Optional Tariffs (OTs), Priority Pricing (PP), Anonymous Mechanisms (AMs), and Franchising (F), are discussed in Section 5.4.

5.1 GENERAL ISSUES

Optimal Regulatory Mechanisms. There has been a considerable amount of work over the last decade or so on the question of optimal regulatory mechanisms.[77] Previous efforts were concentrated on deriving optimal pricing and investment rules for the efficient regulation of a natural monopoly. The application of these rules depended on the regulator having perfect information about demand and cost functions. A regulator interested in maximizing total surplus subject to the firm breaking even (that is, just earning a normal rate of return) would want to impose Ramsey pricing.[78] However, this type of pricing requires information about demand elasticities and marginal costs which the regulator is unlikely to have. Moreover, relying on the firm to provide this information raises the possibility of strategic manipulation.

An optimal regulatory mechanism is designed to maximize social welfare, taking into account informational asymmetries and the incentive of the firm to maximize its profits, not social welfare. The methodology used to solve these problems is principal / agent theory. In this approach, the regulator is the principal who hires an agent, the firm, to act on behalf of the regulator / principal. The agent is hired because it has knowledge and capabilities which the regulator does not. However, achieving the goals of the principal requires effort on the part of the agent and the principal must provide adequate compensation to elicit this effort. The agent will earn information rents and the compensation will exceed the minimum required in a world of perfect information. These mechanisms are Bayesian in that the regulator uses its prior information about the likelihood of possible outcomes to design the regulatory constraint.

As noted by Joskow and Schmalensee (1986, 16), the "flurry of recent theoretical work has so far led to relatively little of practical value." Simple, implementable rules such as 'price equals marginal cost' have not been forthcoming from this literature even with the use of very strong assumptions.

Moreover, the applicability of some of the qualitative results is weakened on two counts. The first is that the focus has almost entirely been on inducing cost-minimizing behaviour in the short-run. Very little has been done on inducing efficient investment behaviour, though

---

77 See Baron (1989) for an understandable introduction to the mathematically formidable literature on optimal regulatory institutions.

78 This involves setting prices such that the deviation between price and marginal cost varies inversely with the elasticity, or price responsiveness, of demand.

some work has been done on minimizing expansion costs.[79] The second is that this literature universally assumes that the regulator can pay subsidies or collect taxes from the firm it regulates. This is typically beyond the powers of most regulatory commissions.[80] Schmalensee (1989) concludes that, by focusing on optimal regulatory institutions, this literature provides little guidance in the design of 'good' regulatory constraints and does not provide guidance to evaluate the performance of regulators.

Instead, most proposed reforms are non-Bayesian. They are not the result of any explicit optimization problem designed to produce an optimal regulatory constraint; rather, they are explicit responses to the alleged incentive problems associated with COS regulation. In particular, most of the suggested alternatives attempt to provide the regulated firm with incentives for cost efficiency. They do this by rewarding the firm through allowing it to earn profits above the allowed rate of return if it reduces costs. This is often done by breaking the link between prices and costs, so that any reduction in costs does not result in an immediate reduction in prices. However an immediate trade-off is suggested. Specifically, if prices do not track costs, the result can be allocative inefficiency.

If cost minimization is the sole objective, this could easily be attained by deregulation. Moreover, any scheme which gives the firm an incentive to reduce costs is also likely to give it an incentive to reduce the quality of service. Finally, any such scheme, since it removes the link between costs and prices, increases the probability of financial distress for the regulated firm. The result may well be an increase in the cost of capital. Given that this cost represents a major element of total costs for gas transmission, the ultimate outcome of incentive regulation in such cases may be an increase, rather than a decrease, in tolls.

Designing Incentive Regulation. Incentive regulation can take many different forms, but the essential feature is that the regulated firm is rewarded financially for achieving standards of performance mandated by the regulator. As noted above, the theoretic literature on optimal regulation may still be lacking in terms of devising optimal schemes which are also practical and implementable. However, it and other research on incentives has provided important insights into the design of applied schemes which can improve overall results by the

79 See Brown, Einhorn, and Vogelsang (1989).

80 If the regulators can use two-part tariffs, the access charge can be used to subsidize a firm.

regulated firm. The resulting rules or guidelines have been outlined by Sappington (1994) and are summarized below.

Asymmetric Information. As noted above, the regulated firm will generally have superior information concerning key relationships such as those relating inputs, output and costs. An important characteristic of a good incentive scheme is that it promotes better use of the firm's superior information about these relationships.

Regulatory Goals. As outlined in Section 3.4, there are a number of regulatory goals or objectives. These involve such things as efficiency, equity, regulatory costs, sustainability, viability, practicality / implementability, and consistency with other policies. It is generally not possible to design an incentive system which will provide improvements with respect to all of these goals. This is similar to the 'Tinbergen rule' which states that if there are, for example, only two instruments, it is not possible to attain simultaneous optimization in terms of more than two objectives. Given this, it is important at the outset to prioritize the regulatory goals and then to design an incentive system which is focused on improvements in terms of only those goals which have the very highest priority.

Sensitivity of Performance Measures. Incentive schemes typically involve mechanisms which change the compensation for the regulated firm based on changes in some measure(s) of performance. If, for example, the objective was to link compensation to the quality of service, it is very important that the performance measure chosen be sensitive to the actions of the firm which are unobserved by the regulator but which the regulator is attempting to influence in order to achieve quality of service objectives. Suppose that the performance measure used for service quality was system outages and the scheme was aimed at encouraging the firm to undertake levels of maintenance appropriate to the quality of service goal. This performance measure would only be suitable if there was a very close relationship between the firm's attention to maintenance and the incidence of service outages. However, if, as would typically be the case with a pipeline, outages occur randomly even when proper maintenance is undertaken, varying the firm's compensation based on service interruptions could mean that the rewards or penalties are largely random. As such, they would not be effective in eliciting the desired responses by the firm concerning maintenance and quality of service.

Variability of Performance Measures. A related issue is the variability or randomness in the relationship between the firm's actions and the results with respect to the performance measure. If the incentive scheme is to be effective in achieving the proper input choices by the firm, it must result in changes in compensation which vary systematically with those choices. Using the previous example, employing service outages as a measure of quality performance for a pipeline could be inefficient because some outages will occur in any given time period independent of the maintenance and other inputs to service quality. Indeed, this could be counterproductive because it would expose the firm to greater risk and this greater risk, through increases in the cost of capital, may mean the costs of inducing the change in behaviour of the regulated firm become greater than the value of any improvements from such behavioural changes.

Limiting Risk and Variability. A general principle in the design of an effective incentive scheme is that it should attempt to hold the firm accountable for outcomes which are clearly under its control but limit the financial responsibility of the firm for outcomes which are mostly beyond its control. This principle can be applied even when the particular performance measure varies according to factors outside the firm's control. For example, the rewards can be made to depend on the source of the variation (so that only those sources under control of the firm are counted), longer-run averages for the performance measure can be used to remove randomness or 'noise,' or 'bands' rather than 'point estimates' can be used in the incentive formulas. The latter approach involves the creation of a 'dead zone' within the bands where rewards or punishment do not vary with changes in performance.

Broad- Versus Narrow-Based Performance Measures. An important issue is whether the scheme should be based on very specific measures of performance, such as 'service response rates,' expenditures on cost-reducing automation, etc., or on general measures, such as average operating expenditures. Except where the regulatory objectives are very narrowly defined, it is generally accepted that broad-based performance measures are strongly preferred. The main reason is that the regulator will not typically have sufficient micro information to pick the specific targets and provide the appropriate level of incentive to induce the firm to behave in an efficient manner. For example, suppose that the specific targets are 'service response rates' and 'reductions in management expense.' The first problem is that, given the particular incentives offered, the firm may not find it economic to alter behaviour with respect to either. A second problem is that, if the incentives are sufficient to change behaviour, the result may be less efficiency rather than more. That is, the incentive may simply result in the firm using more resources to improve service response rates and fewer management resources even though greater efficiency requires, say, alternate use of some equipment, better accounting processes, or more diligent searching for lower cost inputs. A third problem is that the narrow targets will often preclude superior alternatives because the regulator is not aware of every activity, including that associated with emerging new technology, which can achieve the proper overall result of delivering a given level and quality of service at the lowest possible cost.

Performance Benchmarks. Incentive schemes often involve a mechanism to penalize or reward the firm based on some change in the performance measure(s) employed. An important issue concerns the proper benchmark for determining this change in performance. A common approach is to use the status quo or pre-incentive level of performance as the benchmark against which future performance is measured and rewarded. This has two advantages. First, it is important that the firm not be able to alter the benchmark and the status quo performance level would usually meet this test. If the firm were able to affect the benchmark, for example by allowing performance to deteriorate prior to implementation of the incentive scheme, it would be able to obtain rewards simply by returning to typical previous levels of performance. Second, this reduces the risks associated with the implementation of a new regulatory approach since the pre-existing performance would be consistent with the viability of the regulated firm in most cases.

A further rule is that the benchmark should be consistent over time. This is important to prevent strategic manipulation such as in the form of 'cycling' behaviour. In these cases, there is an incentive to shift costs from one year to another in order to maximize rewards. For example, if the firm were provided with a reward for reducing operating costs from one year to the next, with no penalty if operating costs increase, it would be optimal for the firm to delay some costs in year one and receive a reward, incur the higher costs in year two and then shift costs forward again in year three to receive another reward.

Optional Incentive Plans. It is generally preferable to offer two or more distinct incentive plans and then allow the firm to select one of the plans. One plan might set a modest goal for cost reduction and provide only small variations in the compensation to the firm from changes in performance. Another plan would have a more ambitious cost reduction target and provide larger rewards for achieving the target and larger penalties for not achieving it. Such an option approach takes advantage of the superior knowledge that the regulated firm has concerning the scope, capabilities and costs of activities aimed at improving efficiency and reducing costs. As demonstrated by Sappington (1994, 258–261), relative to the case where the firm is not presented with options, optional incentive schemes are more likely to result in 'win-win' situations, efficient outcomes and general acceptance.

Regulatory Commitment. A common assumption in the literature on incentive schemes is that the regulator has the ability to execute any commitments embodied in the particular scheme adopted. Without such a commitment, the incentive effects will be reduced because the firm will be less likely to undertake the expenditures and other efforts required to raise the likelihood of good performance. However, as a practical matter, the ability of regulators to make longer term commitments is limited. They cannot usually bind their successors and, also, political shifts and policy changes by governments can often require alterations in the regulatory regime. This means that, in practice, incentive schemes which require strong, long-term commitments by the regulator may not be feasible. In summary, Sappington (1994, 263) suggests that, in selecting an incentive program, regulators should take care "... to explicitly promise *ex ante* only what can be delivered *ex post.*"

Plan Duration and Flexibility. It is not possible to design an incentive scheme that is compatible with all contingencies, many of which are unforeseeable. In the case of an unforeseen event it may be desirable to make adjustments if it has a significant impact on regulatory performance. That is, some flexibility is optimal but it is often unclear how much flexibility is optimal. If there is too little, the viability of the firm may be threatened or it will receive unwarranted or unacceptable rewards. If there is too much flexibility, the incentives to engage in efficient planning and cost-reducing activities will be greatly reduced.

Sappington (1994, 268) argues that some limited flexibility to make after-the-fact adjustments is usually appropriate. This should consist of, first, a clearly defined evaluation or adjustment point which is spelled out in the incentive scheme. For example, it might be specified that the scheme is to be evaluated after it is in place for five years. This would provide time for the incentives to become integrated in the firm's decision-making and operations, but it would not be so long that errors or shortcomings in the scheme will cause large, unanticipated deviations. Second, the incentive regulation should also specify the events which will trigger changes in the scheme over the period from implementation to the first review. These events would typically be limited to those which are quite significant, which could not be reasonably anticipated and which are largely beyond the control of the firm.

5.2 PRICE LEVEL INCENTIVE SCHEMES

There are four main incentive mechanisms which focus on the regulation of prices charged by the firm rather than on its profits. These are price caps (PCs), automatic rate adjustment mechanisms (ARAMS), sliding scale plans (SSPs), and yardstick competition (YC). They all involve placing ceilings or upper limits on prices or tolls. Since prices become independent of costs, the firm has greater incentives to reduce costs. This is because it is allowed to keep, at least until the next review of the scheme, the higher profits resulting from cost savings. These price level schemes differ mainly in how prices are adjusted to reflect changes in demand and / or cost conditions. In some sense, they all involve caps on prices. However, by convention, only one of these approaches is formally referred to as a price cap scheme.

Price Caps (PCs). These were adopted in a number of jurisdictions in the 1980s and represent an institutional replacement for COS regulation. For example, British Telecom, British Gas, and American Telephone and Telegraph Company (AT&T) were all put under a price cap constraint. Oil pipelines in the U.S. were scheduled for PC regulation as of January 1, 1995. Further, a proposal by a group of large shippers involves similar PC regulation for Canadian oil and gas pipelines and a negotiated price cap scheme (with profit sharing) for Interprovincial Pipe Line Inc. (IPL) has recently been approved by the NEB.[81]

Fundamental Characteristics. Acton and Vogelsang (1989) define four characteristics which are fundamental to price cap regulation. These are:

(i) A price ceiling is imposed by the regulator. The firm can charge any price at or below the ceiling.

(ii) Price ceilings do not necessarily have to apply to a single product. Instead they can be defined for baskets of goods or services. The price of each basket of goods / services is a constructed price index and each basket has a ceiling on its price index. If the price ceilings apply to baskets rather than to each of the individual commodities, the firm operating under PC regulation has the ability to alter relative prices.

---

81 See Imperial Oil Limited (1994) for the general price cap proposal. The incentive scheme for IPL is described in Section 6.1.

(iii) The price ceilings are adjusted periodically by a pre-announced and exogenous (to the regulated firm) adjustment factor. This factor determines by how much rates can rise (or fall) in any period. The factor is usually comprised of two components. The first component is supposed to pass cost increases (or decreases) due to changes in the prices of inputs through to final prices. The second measure is a productivity offset. The price ceiling is reduced by a percentage equal to the productivity growth offset. This component is designed to allow consumers (shippers) to share in any cost efficiencies attained by the firm.

(iv) The adjustment factors, baskets, and indices are periodically reviewed and updated.

Symbolically, the relationship between the maximum price in period $t\left(p_t\right)$ and the price in the previous period $\left(p_{t-1}\right)$ for a single product firm is:

$$p_t = p_{t-1}\left(1 + RPI_{t-1} - X\right) \tag{5.2.1}$$

where RPI_{t-1} is the percentage change in the price index during period *t-1* and X the productivity offset.

The Firm's Responsibilities. Under this regulatory scheme the price(s) the firm is allowed to charge are constrained by the price cap, equation (5.2.1). The firm is not permitted to charge prices in excess of the ceilings in a given period. However, it could charge lower prices and, if the price caps are set on a basket of goods or services, it can adjust relative prices, provided the cap on the basket is not exceeded. This means that it can raise some prices, provided others fall to compensate. Moreover, the firm is required to satisfy demand at the prices it eventually selects. Subject to this constraint on its prices and satisfying demand, the firm determines its inputs and the technique of production. The cost side of its operations is essentially deregulated. In particular, the firm determines the size and timing of capacity expansions unless the base prices are reset as part of a review by the regulator for any significant expansions.

The income of the firm will depend on the following factors:

(i) The relationship between the initial prices and costs.

(ii) How closely changes in the price index match the changes in its actual costs.

(iii) How closely X reflects actual cost reductions arising from efforts at cost minimization, technological change, depreciation and any decreasing costs associated with volume expansions.

(iv) Its ability to rebalance rates and offer new services.

(v) The effect of the change in regulation on its risk and, hence, on its cost of capital.

(vi) The nature of the periodic reviews and whether or not profit data will be used to reassess the price cap.

(vii) Whether the regulator can commit to not reassess the price cap between reviews and whether the regulator still monitors profits between reviews.

The Regulator's Responsibilities. Implementation of a PC regime requires that the regulator establish the regulatory constraint by defining the variables which comprise equation (5.2.1). These variables and a discussion of the options are presented below.

(i) *Initial Prices.* The initial price for each product or service must be set. If prices are set too low, the financial viability of the firm is threatened. On the other hand, prices set too high initially means that any monopoly power of the firm is insufficiently restricted.

The regulator has two options. The first is to use the prevailing COS tolls. These have an advantage in that the regulator knows that the firm is viable at these levels, demand can be met at these tolls, and, presumably, the tolls are not excessive. However, there may be problems with using COS tolls. These tolls are based on historic cost accounting conventions and, as such, may not reflect true economic costs. In the absence of regular additions to capacity, the current treatment of depreciation under COS regulation results in tolls which will exhibit front-end load patterns. As illustrated earlier in Figure 4.1, tolls generally exceed economic costs initially and then, in later years, they can be significantly less than the economic cost. However, if there are frequent expansions and rolled-in tolling methodology is employed, the resulting tolls may approximate economic costs.

Except in the latter case, use of existing tolls to set initial prices under a PC scheme means that either the shippers or the pipeline company will receive a windfall gain since the switch to price-cap regulation locks in the depreciation of previous years. If the accumulated depreciation has been too small (relative to true or economic depreciation), the pipeline's financial viability will be threatened. If it has been too large, the price-caps will begin with tolls well in excess of marginal cost.

The regulator could adjust tolls away from the levels determined under COS. Usually, movements away from COS tolls would be designed to move prices closer to marginal costs and eliminate or reduce cross-subsidies. If prices are adjusted to reflect economic costs, the direction of adjustment will depend on the age of the facilities. If the facilities are relatively new, the adjusted prices would fall. If the facilities are heavily depreciated, the adjusted prices will rise. These types of price or toll adjustments can be very significant and may arouse considerable political opposition.

(ii) *The Commodity Baskets.* The regulator must decide which services will be grouped together and, therefore, be subject to the same price ceiling. In selecting whether or not services will be bundled and, if so, which services go in which bundles, the regulator determines the extent to which the firm will have flexibility to change relative prices. Applying a price ceiling to a price index based on the individual prices of two or more separate services allows the firm the opportunity to adjust relative prices.

To the extent that relative prices under COS methodology reflect equity and / or political considerations, the use of a basket of services provides the firm with the opportunity and the flexibility to engage in profit-increasing price discrimination. This means that it will find it profitable to raise the prices of those services whose demand is inelastic and lower the prices of those services which have a more elastic (price responsive) demand. This is a movement in the direction towards Ramsey pricing and may be welfare-improving. It certainly will allow higher profits but political opposition to this type of rate re-balancing should be anticipated.

Most tolling systems for gas pipelines involve two-part tariffs. That is, they consist of a fixed access or demand charge and a variable or commodity charge which reflects the utilization of the reserved capacity. Unfortunately, very little work has been done on reconciling two-part tariff structures with price-caps. An important issue is whether demand (access) charges and commodity (volumetric) charges are to be treated independently or are to be aggregated into a price index. Alternatively, the question is whether the demand charge is to be converted into a per-unit toll on the basis of an exogenously specified load factor.

The regulator will also have to determine the regulatory treatment of the introduction of new products and services. For example, it will be important to specify how the introduction of new products and services will be incorporated into the price cap mechanism. One possibility is to add them to existing baskets and construct new price indexes. Another approach might be through substitution so that some other services are eliminated from the baskets.

(iii) *The Price Index.* A mechanism is required to adjust the price caps to refect changes in costs and demands over time. In order to increase the incentives for cost efficiency, the price index should reflect changes in costs and demand which are beyond the control of the firm. That is, the objective is to break the link between the firm's prices and its costs. However, breaking this link may threaten the viability of the firm if its costs, due to factors beyond its control, rise faster than the price index (net of the productivity or X factor) used for the price cap.

Using a price index which reflects the actual costs of the firm provides it with the opportunity to strategically manipulate its prices by not minimizing costs: in such a case the formula will essentially institute cost-based prices. This rules out choosing an index based on the costs of the regulated firm. As a consequence of the need to adjust prices to reflect changing costs without eliminating the incentives for cost reduction, the indexes adopted in other jurisdictions or recommended by proponents of price cap regulation are indexes of economy-wide inflation or price increases. Possible economy-wide inflation indexes include the Consumer Price Index (CPI), the Producer Price Index (PPI), or the Gross Domestic Product Price Index (GDP-PI).

(iv) *Productivity Adjustment Factor*. The inclusion of an X factor is a means of passing the anticipated cost savings on to customers through lower prices. The X factor represents *expected* productivity gains. The question of how it is to be determined is particularly vexing and important. The historical pattern of technological change is almost certainly irrelevant. It is also important to note that the factor chosen is supposed to be fixed between reviews. Too high a value for X places the firm in jeopardy; if it is too low the firm is the main beneficiary of cost reductions and prices do not reflect costs (that is, it leads to allocative inefficiency). Further, as discussed in Section 4.3, the long-term trend in real costs for pipelines will not be dominated by productivity gains. Rather, the key determinants will be depreciation and the resulting front-end load toll patterns, in combination with changes in inflation and capacity additions. Consequently, the X factor in such cases cannot be set simply on the basis of expected productivity changes.

(v) *Periodic Reviews.* In the long-run, price adjustments to reflect changes in demand and costs are necessary given the likely inability of the price index to track the actual costs of the firm and the inability of the regulator and the firm to accurately predict the values for X. These adjustments are achieved by a comprehensive review of the price cap. The regulator must decide the frequency of a periodic review and the criteria for review and reassessment of the components of a price cap. An important difficulty with such a review is the likely inability of the regulator to completely avoid using information on profits to revise price ceilings and the value of X. As a consequence the new price ceilings will likely be based on costs (assuming the base is reset using COS methodology) and the value of X will likely be revised based on profit levels achieved since the last review.

The implication that a review will result in the re-establishment of cost-based prices means that the length of time between reviews becomes important. Too short a time between reviews means a reduction in the incentives for cost minimization and reduction. Too long a time between reviews means that prices and costs are likely to diverge significantly. In addition, the regulator will have to determine whether or not the time between reviews is fixed or whether there will be recourse to reviews under special circumstances. If the latter applies, it will be important to specify what constitutes special circumstances. It is also important for the regulator to indicate whether the profits of the firm will be monitored between reviews and whether rates of return which are higher than expected will trigger a review.

Automatic Rate Adjustment Mechanisms (ARAMS). Automatic rate adjustment mechanisms are very similar to price caps. The basic approach involves adjusting prices automatically as costs change. This is usually done by using a formula which relates output prices to input prices.[82] ARAMS can be based on changes in the firm's own costs or on industry costs; they can allow for either full or partial cost pass-through;

---

82 As Brown, Einhorn, and Vogelsang (1989, 37) note, ARAMS can be based on input prices, input quantities, or both. Here we consider only input prices. Comprehensive measures based on input quantities require measurement of total factor productivity, something which is notoriously difficult to do at the required degree of precision.

and, they can be comprehensive or they can be based only on the costs of certain inputs.

The three schemes which have attracted the most attention are strict formula pricing (comprehensive own-cost), restricted own-cost, and comprehensive exogenous cost.

(i) *Formula Pricing.* Using a formula based on the actual costs incurred by the firm is a form of cost-plus pricing. An initial price is specified and, as input prices change, output prices adjust according to the formula. This type of regime is very similar to that outlined in the discussion of price caps. The essential difference is that, in the case of an ARAMS, the price ceiling adjusts by means of a detailed formula which tracks changes in a weighted average of input prices. The weights are set based on the share of total costs accounted for by each factor of production. In some cases, such as the formulas used to adjust producer milk prices, other factors have been included to capture shifts in the level of demand.[83] In contrast, PC schemes employ indexes which capture economy-wide price changes (and sector-specific productivity change) rather than changes in the firm's actual costs.

(ii) *Restricted Own-Cost.* This type of adjustment formula is designed to reduce the frequency of rate hearings where a high frequency is due to the volatility of certain input prices. Cost changes due to input price changes for key inputs are, via a formula, passed through automatically into changes in output prices. Purchased gas adjustment clauses and automatic fuel adjustment clauses are two examples of this type of scheme.

(iii) *Exogenous Cost Index.* In this formulation, the automatic rate adjustment mechanism is used as the price index in price cap regulation. The adjustment to the allowed prices of the firm, or the cost pass-through, is a weighted average of input price changes, where the weights are the input shares in total cost. However, unlike formula pricing, average industry prices, or some other comparable price index which is exogenous to the firm, is used instead of an index of actual input prices paid by the firm.

Sliding Scale Plans (SSPs). Under these schemes, efficiency gains are shared between the firm and its customers. Sliding scale plans are usually incorporated into PC regulation (this combination will be referred to as a PC-SS scheme). The Alberta Petroleum Marketing

83 See Mansell, Wright, and Kerr (1984) for a discussion of these and other pricing indexes.

Commission (1991) suggested that PC-SS schemes might be applicable to gas transmission in Canada. Braeutigam and Panzar (1993, 196) report that, *circa* mid-1992, just under half of the lower 48 states in the U.S. used PC-SS schemes to regulate local telephone companies (also known as 'The Baby Bells').

A PC-SS regulatory regime is identical to a PC regime as discussed above, except for how the price-ceiling is adjusted. Under a profit-sharing or sliding scale plan, prices are adjusted based on the realized rate of return. Following Brown, Einhorn, Vogelsang (1989, 45), generic sliding scale plans can be represented by:

$$p_t = p_{t-1} + \alpha(s^* - s_{t-1})\frac{K_{t-1}}{q_{t-1}} \tag{5.2.2}$$

where α is the sharing parameter, $0 \leq \alpha \leq 1$, s^* is the target rate of return, s_{t-1} is the actual rate of return in the last period, K_{t-1} is the rate base last period, p_t is price in period t and q_{t-1} represents the sales in the last period. If α equals 0, this scheme is equivalent to a rigid price cap, while α equal to 1 corresponds to COS regulation.

The representation in equation (5.2.2) assumes that the price decrease from exceeding the target rate of return is linear in the efficiency gains ($s^* - s_{t-1}$). In practice, however, implementations of PC-SS plans differ from this in that there is a level above s^* (denoted $\bar{s}$) where, for higher rates of return, the sharing parameter equals one and, for rates of return less than s^*, the price does not change. If the rate of return is substantially below s^*, a rate hearing is held. Finally, instead of adjusting prices in the future, actual implementations often involve refunds.

Critical to implementing the PC-SS scheme is the determination of the sharing parameter, α, and the upper-bound on the rate of return, $\bar{s}$, by the regulator. These parameters determine the return to the firm from cost-minimizing behaviour; the larger they are, the more likely the regulated firm will have incentives that match those of an unregulated firm.

Yardstick Competition (YC). The idea behind yardstick competition is based on one aspect of how a competitive market works. In such a market, competitors' costs determine the prices that can be charged by the firm. A strong yardstick approach would be to set the toll on one pipeline equal to the average cost of all the remaining pipelines. Since the firm's prices are not based strictly on its costs, the firm has a stronger incentive to minimize its costs. Moreover, in a dynamic context, lower costs today translate into lower prices in the future. Once again, except for how the price ceiling is set and adjusted, the mechanism is essentially a PC regime.

5.3 PROFIT LEVEL INCENTIVE SCHEMES

These schemes focus on changing the profit rates of the regulated firm in response to changes in some measure of performance. The four main approaches are profit sharing (PS), banded rates of return (BROR), benchmarking (BM), and capital cost incentives (CCIs). In most cases, these are really potential modifications to traditional COS regulation rather than stand-alone regulatory approaches. That is, for the most part, they are an attempt to increase the incentives for efficiency under COS regulation.

Profit Sharing (PS). This regime is a variant of traditional profit level regulation. For it to work, the regulator must be able to make commitments. Under this plan, cost reductions and increased efficiencies, as measured by increases in the profits of the firm, are shared between customers and the firm permanently. For instance, suppose that the firm had been authorized to earn a 10 percent rate of return but at the next rate hearing it was determined that the realized rate of return was 15 percent. The difference is assumed to be due to increased efficiency and cost savings on the part of the firm. The firm is allowed to keep some percentage of this. At the rate hearing, prices are adjusted downwards to reflect the share of profits awarded to consumers. Otherwise this scheme is identical to COS regulation. It might also be noted that traditional COS regulation represents a special case of PS where all of the extra profits arising from cost reductions between rate hearings are maintained by the firm but, at the next rate hearing, prices are adjusted downward to completely eliminate the extra profits so that all the benefits of the efficiency gains are passed along to consumers. In other words, after the initial period of higher profits, the customers of the firm are awarded all of the gains in efficiency through downward price adjustments. As described in Section 6.1, the incentive scheme recently applied to IPL includes a profit sharing mechanism in combination with a price cap.

To implement a PS scheme, it is necessary that the regulator can credibly commit to allow prices at later reviews to be set at levels above the firm's costs (where those costs would include only a 'normal' rate of return on investment). If such a commitment cannot be made, it is unreasonable to expect the firm to change behaviour based on anticipation of higher longer-run rates of return (relative to the case of traditional COS regulation).

Banded Rate of Return (BROR). This is also a variant of COS regulation. It involves determining a lower and upper limit around the authorized rate of return. If the realized rate of return is within this deadband, there is no regulatory action: the firm bears the loss (in the sense that it achieves a rate of return less than the authorized rate of return) or it keeps the extra profit. A version of this type of regulation is used by the NEB to regulate oil pipelines in Canada. In that case, the deadband is plus or minus two percentage points of the authorized rate of return.

Usually with this approach, the authorized rates are left unchanged until the bounds are exceeded. When the firm's rate of return goes outside of the bounds, rates (tolls) are adjusted either through a sliding-scale scheme or by holding a rate hearing.

Benchmarking (BM). In a benchmark scheme, the allowed rate of return depends on the firm realizing some operational goals. This approach represented the core of a proposal made by the Canadian Association of Petroleum Producers (CAPP, 1991) in the NEB's Workshop on Incentive Regulation. CAPP raised the issue of developing operational performance objectives and linking the allowed return on equity of the pipeline to these performance objectives. One interpretation is that the proponents would like to see a formula developed whereby the allowed rate of return of a pipeline is built up from an assessment of how well the firm did relative to about twenty performance indicators.

Many regulatory commissions in the United States have introduced narrow incentive payment programs based upon benchmarking using a specific measure of firm performance. The Edison Electric Institute (1987, 11) defines such an incentive program as one which "(*a*) is intended to improve regulated utilities' performance, (*b*) evaluates utility performance against specific, pre-defined standards, and (*c*) provides incentives (rewards) or disincentives (punishments), depending on the utility's performance in relation to applicable standards,..." (as quoted in Berg and Jeong 1991, 45).

The appeal of such schemes is obvious. Once standards are set, it is often easy to determine whether or not they have been met. Moreover, this allows the regulator to improve the performance of the utility in areas with high visibility. A recent survey in Public Utilities Fortnightly (1991b, 49–60) of the Chairmen of State Regulatory Commissions suggests that these programs work, at least in terms of achieving the narrow, pre-defined objectives. However, as explained in Section 6, they are less successful in achieving true efficiency as reflected by the overall level of prices or tolls.

Capital Cost Incentives (CCIs). The strong influence of the rate base on tolls under COS regulation has led some observers to suggest specific performance indicators to control construction or expansion costs. An incentive scheme to control expansion costs should have two properties. First, it should induce the firm to provide the firm's best estimate of the investment costs. Secondly, it should provide the firm with incentives to minimize these costs. One such approach is suggested by Brown, Einhorn, and Vogelsang (1991, 335). This is a modification of a scheme proposed by Laffont and Tirole (1986). Under this approach, additions to the rate base do not equal actual investment costs. Instead, the firm is provided with incentives to minimize costs and provide accurate estimates by making the rate base addition a function of the cost estimate and a function of the difference between actual costs and the original estimate. The rule used to determine rate base additions is:

$$B = A + x(E) + y(E)[E - A]\ , \tag{5.3.1}$$

where B is the allowed addition to the rate base, A the actual cost of the expansion, and E is the predicted costs. The two parameters, x and y, determine the amount of the original estimate which is included in the rate base and the share of any cost overrun (underrun) which is deducted (added), respectively. These shares depend on the initial estimate provided by the firm. The higher the estimated cost, the smaller are x and y. The firm has an incentive to minimize construction costs since any reduction in costs below the original estimate increases its rate base. However, this also gives it an incentive to overestimate. This incentive is controlled by making x and y decreasing functions of E.

If $y = 1$ and $x = C - E$, with C a constant, this scheme becomes a fixed price contract. Independent of its costs, the firm is allowed to increase its rate base by C. The firm, therefore, has an incentive to minimize construction costs. However, the firm carries all of the investment risk associated with cost overruns. If $y = 0$ and $x = 0$, this scheme allows complete pass-through. It obviously reduces the incentives for the firm to minimize costs and the firm does not face any risk for cost overruns. Under this scheme, large values for $y(E)$ combined with smaller values for $x(E)$ assign more of the risks of the investment to the firm and this provides it with stronger incentives to minimize capital costs.

With such a scheme, the firm will have incentives to minimize construction costs and to make accurate cost estimates prior to the beginning of construction. Accurate cost estimates in the planning stage should result in better decisions concerning the timing of capacity expansions. This incentive mechanism is similar to *ex post* prudence reviews. However, instead of disallowing cost overruns after the fact, it partially disallows them *ex ante*.

A variant of this approach involves rewarding the firm if it meets its cost estimate by allowing it to earn a higher rate of return, rather than allowing an extra addition to the rate base. The advantage of the scheme proposed above is that it provides better incentives for the firm to truthfully reveal its best estimate of construction costs.

Yet another (related) approach is the use of Expedited Certificates. Here, the regulated firm is given the flexibility to construct new facilities, but they would be initially tolled on a stand-alone or incremental basis.[84] At a subsequent hearing, usually two or more years after completion, the regulator would make a decision as to the proportion of these expansion facilities that would be allowed to be rolled into the existing rate base. With this approach there is obviously a strong incentive to ensure that the construction costs of the new facilities are minimized given that there is the possibility that none or only part of the costs would eventually be allowed into the existing rate base.

5.4 OTHER APPROACHES

The eight schemes outlined above constitute the core of the most commonly proposed approaches to incentive regulation for utilities. However, there are other possibilities, at least at a theoretic level. These include deregulation of pipelines, franchising, the creation of competitive secondary markets for capacity, and various schemes aimed at increasing pricing efficiency. The basic elements of each of these alternatives are outlined below, along with a brief assessment of their applicability to the case of major gas pipelines in Canada.

Deregulation and Supply-Basin Competition. Some observers have argued that, since the ex-Alberta pipelines compete at the city gate with other supply basins, there is no strong justification for public utility-type regulation of these systems. Another argument sometimes presented is that the ex-Alberta pipelines essentially compete with each other to undertake the next expansion or build the next new pipeline.

These arguments have already been discussed in Section 3.2. It will suffice to say that, while there is some merit to at least the argument that there is competition for new capacity, it is unclear that there would be enough competition in most cases in Canada to warrant the deregulation of tolls.

---

84 Given the hold-up problem discussed in Section 2, the firm would usually only proceed if a sufficient number of shippers had made a commitment in the form of a long-term transportation contract.

Franchising. The idea of having firms compete 'for the market' instead of competing 'in the market' goes back to Demsetz (1968). The regulator auctions the right to provide service in the market. This right goes to the firm that submits the lowest price. Demsetz argued that at least for a single product firm, this would result in the most efficient pricing subject to the constraint that the return to the firm be normal; that is, it would result in Ramsey pricing. In a static world, this would be an attractive solution to the problem of a single-product natural monopoly. If the natural monopolist is a multiproduct firm, it is no longer clear what submitting the lowest price means. Moreover, when the bid is multi-dimensional, for instance involving price and reliability, determination of the optimal bid is not obvious.

Williamson (1976), in a study of the application of auctioning of cable television franchises in the United States, documents the difficulties that arose from this scheme in a world where costs and demands change. Changes in costs and demands require a change in prices. The equivalent of a rate hearing evolved to determine how prices would be allowed to change.

Alternatively, the regulator could periodically re-open bidding for the franchise. In these circumstances the franchises would be short-term. However, this is really only workable if there are no sunk costs. If there are sunk costs, the equivalent of a rate hearing must be held to determine what the rate base is for a potential transfer between the old and new franchisee. As a result franchising evolves into something very similar to COS regulation.

One variant that has been suggested is that there be open bidding for all aspects of facilities construction, including design and interim financing, with the possibility of ownership remaining with other companies, including those of shippers. The regulated pipeline company would still be the operator, but would not necessarily be in charge of constructing and owning all of the links in the network. Such an option would be compatible with all of the regulatory regimes described above. However, several points should be highlighted.

(i) Such arrangements already exist. For example, the gas transmission system in Alberta already includes non-NOVA transmission facilities which NOVA uses and the payments for this use represents part of NOVA's revenue requirement.

(ii) If the main objective is to ensure 'harder bargaining' with respect to construction costs, comparisons of the regulated firm's costs of constructing facilities and the costs of the same facilities if constructed by another party can be made. The results of such exercises can then be incorporated in the traditional regulatory regimes, as well as in many of the new-style incentive regimes.

(iii) If used on a significant or ongoing basis, there is likely to be a conflict between the operator and the builder / owner. This conflict arises because actions taken by the builder / owner will affect the costs of operation and the options of the operator. However, since the nature of their respective involvements is by nature sequential, it is the operator who is left to sort out any lingering difficulties associated with construction. It would not be realistic to expect that all of these difficulties would or could be covered under contractual remedies.

(iv) To the extent that the operator sets standards and oversees construction, the regulated firm may have a strategic advantage over other bidders. Further, in most cases only the regulated operating firm would have the information concerning system-wide issues to be able to optimally design, plan and construct expansion facilities.

Secondary Markets. There has been considerable interest in both the regulatory arena and in the regulation literature in the creation of complete markets for secondary capacity (Alger and Toman, 1990).[85] The National Energy Board in Canada recently enacted regulations which allow for an unrestricted market in secondary capacity. A complete secondary capacity market is a market in which shippers can re-sell their rights to existing capacity at a market determined price.[86] However, the original contractual obligation, such as the obligation to pay the demand charges for an agreed period of time, would remain intact. Consequently, the investment by the pipeline remains protected. For example, if the contracts underpinning an expansion carried an obligation to pay $1 million per period but the best price in the secondary market for that capacity amounted to $600,000 per period, plus payment of the commodity or variable charges, the original party who signed the contract would still be required to make up the difference of $400,000 per period.

Proponents of unrestricted or complete secondary markets argue that such a scheme would result in efficient rationing (those who value

85 See National Energy Board (1994a) for a discussion of this issue in a Canadian context.

86 This differs from the informal, very limited 'markets' allowed by regulators in the past which mainly involved setting the rules for the temporary or permanent reassignment of capacity rights.

capacity the most are able to purchase it from those who value it less) and that the prices in these markets would be reasonably accurate indicators of the value of new capacity to shippers.[87] It will suffice to note here that this option is compatible with all of the regulatory regimes set out in earlier sections and probably deserves serious consideration. However, there are some important issues in the creation of complete secondary markets that should be noted.

(i) While secondary markets do allocate capacity efficiently when capacity is constrained, the scarcity rents may not accrue to the pipeline but to the holder of the capacity rights. This raises both a distributional and an efficiency issue. It is an efficiency issue because the pipeline is often a natural monopoly. Diverting the rents to capacity holders means that tolls must deviate even further from marginal costs, thus creating a greater distortion, in order to meet the pipeline's revenue requirements. Transferring these rents back to the pipeline, as is currently done with interruptible proceeds, can reduce the price distortions by lowering the tolls for firm service closer to marginal cost. The distributional issue is whether any rents associated with capacity rights should accrue to those who happened to obtain these rights at costs significantly below market value or whether they should be distributed more broadly by lowering tolls for all shippers.

(ii) Large discrepancies between the regulated tolls and the secondary market capacity prices may result in non-optimal investment decisions. For example, as explained in Section 4.3, suppose that incremental tolling is used for expansions. Given front-end loading, the toll in the early years would almost certainly be well above the price which the capacity would fetch in the secondary market. This could result in under-investment. Potential shippers would be reluctant to sign the long-term contracts necessary to underpin construction because of the higher risk. In addition to the normal risk of demand charge losses, there would be additional risk because if any of the new capacity was traded on the secondary market, the lower capacity prices in that market would give a competitive advantage to those who did not sign the long-term contracts associated with the new capacity. This would lead to displacement of the original shipper from the market and almost

87 For example, if the capacity on a certain system consistently traded at prices in excess of the marginal or incremental cost of expansion this would provide an efficient market signal that additional capacity should be constructed.

certain demand charge losses by the original shipper who signed the long-term transportation contracts.

(iii) The transaction costs associated with independent secondary markets for connected but separate pipelines may be high. An integrated market would reduce transaction costs and increase efficiency.

(iv) There may be a concern that large groups, such as supply or demand aggregators, could exert considerable market power in the short run by monopolizing capacity rights. Teece (1990) argues that there are likely to be large economies associated with aggregation. The larger the aggregator, the easier it is to satisfy random demand with random supplies and, consequently, it is more efficient to have relatively few aggregators on a pipeline corridor. If this view is correct, one of the consequences of a secondary market in capacity will be the accumulation of capacity rights by large aggregators and these aggregators may acquire the monopoly power associated with limited pipeline capacity. The result could be lower field-gate prices and higher city-gate prices.

Priority pricing is an alternative to complete secondary markets for capacity. It may provide the advantage of lower transaction costs and overall tolls (since it would transfer the scarcity rents associated with capacity in times of shortage back to the pipeline for the benefit of all shippers). Priority pricing is discussed below.

Pricing Efficiency. Allocative or pricing inefficiencies are often associated with traditional regulatory regimes, but need not be. For example, pricing inefficiency is not necessarily an attribute of COS regulation. Rather, it often stems from the use of fully distributed cost (FDC) pricing for a firm providing multiple goods or services.

In the case of a natural monopoly where marginal cost is below average cost, efficient prices have two properties. The first is that they minimize the social cost (reduction in surplus) associated with raising prices above marginal cost in order to cover total costs or, alternatively, they minimize the social costs of meeting the revenue constraint. Price must be raised above marginal cost when there are economies of scale / scope or the firm will earn negative economic profits. The second property is that prices should be flexible enough to respond to changes in market conditions. In the context of pipelines this can be very important since gas transmission is non-storable. Efficient prices would respond to random changes in demand and capacity.

This section outlines a number of proposals that have been advanced as replacements for fully distributed cost pricing. These include the deregulation of relative prices, optional tariffs, priority pricing, and efficient pricing mechanisms.

Deregulation of Relative Prices. Allocative or pricing inefficiencies are a potentially significant problem associated with COS regulation, especially when fully distributed cost (FDC) pricing is employed. It will be recalled from the discussion in Section 3.3 that this arises from deviations between prices and marginal cost and from the inability to set individual prices for the regulated multiproduct (or multiservice) firm to reflect the relative elasticities in various markets.

One proposal to improve the efficiency of the rate structure of a regulated multiproduct firm is to deregulate relative prices. From an initial set of prices at time t (p_i^t), the firm is allowed to charge any prices in $t+1$ (p_i^{t+1}) as long as the following constraint is satisfied:

$$\sum_{i=1}^{n} p_i^{t+1} x_i^t \leq \sum_{i=1}^{n} p_i^t x_i^t \qquad (5.4.1)$$

where x_i^t is the quantity of commodity i sold in period t and n is the number of products or services supplied. The constraint requires that the bundle of goods and services purchased in period t costs less in period $t+1$ than in the previous period.

This constraint can be rewritten as:

$$\frac{\sum_{i=1}^{n} p_i^{t+1} x_i^t}{\sum_{i=1}^{n} p_i^t x_i^t} \leq 1 \qquad (5.4.2)$$

The initial prices are determined by the regulator, so they are likely to be fully distributed cost (FDC) prices.

Equation (5.4.2) is a Laspeyres price index.[88] The denominator is the firm's revenue in period t and the numerator is the firm's *pseudorevenue* in period $t+1$. Pseudorevenue is the revenue the firm would receive if it sold last period quantities at current period prices.

Laspeyres price indexes have attractive efficiency properties. With deregulated relative prices, firms can increase their profits by altering relative prices to reflect costs and demand elasticities. Provided this is done subject to a constraint on average prices based on a Laspeyres

88 A Laspeyres price index is one for which the basket of goods used to weight prices is based on actual purchases in the past or the base period.

price index, consumers in aggregate will also benefit: in other words, consumer surplus will rise.

To intuitively understand this result, suppose that there is only one consumer. The fact that the index must decline or remain constant means that this consumer will be able to buy the same basket of goods and services as in the last period at new prices which make the total expenditure the same or less. If they choose to buy different quantities, they must be better off.

When there are many consumers in the market, constraining prices via a non-increasing Laspeyres price index means that reductions in consumer surplus from price increases will be equal to or less than increases in consumers' surplus from price decreases. Thus, deregulating the relative price structure can increase both profits and consumer surplus and, as a result, total surplus rises. Brennan (1989) shows that the relative price structure of a firm subject to a Laspeyres price constraint in every period will eventually converge to a relative rate structure similar to that with Ramsey prices.[89] This means that the relative prices will be the same as Ramsey prices, but the profits of the firm need not equal zero.

There are two main difficulties with full deregulation of relative prices. The first is an equity issue. The second is the incentive for the firm to behave strategically.

Deregulation of the relative prices of goods and services which have markedly different price elasticities will undoubtedly result in political controversy. Movement by the firm to a more efficient pricing structure is not likely to be welfare-improving for all customers. Many applications of fully distributed cost pricing implicitly involve cross-subsidization of consumers with inelastic demands by those with elastic demands and movements towards efficient pricing will eliminate this subsidy. While the profits of the firm and the surplus of consumers whose demand is price-sensitive or elastic will increase, the surplus of consumers whose demand is inelastic will fall. In the typical case, this occurs because the movement to efficient prices requires a decline in the prices charged elastic consumers and an increase in the price charged inelastic consumers.

In the context of pipelines, an extreme example might arise when a profit-maximizing pipeline adjusts the relative tolls for firm and interruptible service. A profit-maximizing firm free to set relative prices may discount interruptible rates and raise the rates for firm service. Even if

89 See also Bradley and Price (1988) and Vogelsang (1989).

the pipeline can only adjust tolls for firm service, it may increase rates for elastic markets and decrease rates for inelastic ones. The greater the variation, the larger the adjustment in relative prices. Achieving an optimal balance between price adjustments and protecting the interests of consumers with inelastic demands is likely to be a very difficult problem.

Brennan (1989) has shown how a firm in an environment of changing costs and / or demands can act in a strategic manner to increase its profits and reduce consumer welfare when it is allowed the flexibility to determine the relative price structure subject only to a Laspeyres price index.

Shifts in the demand curve raise the question as to what quantities should be used to weight the prices of last period. In order to correctly credit the firm for both the welfare gains and losses created when it changes its prices, price changes need to weighted by "what last period's quantities would have been had the current period's demand curve been in effect." (Brennan 1989, 139) As Brennan points out, predicting current demand at last period's prices in order to generate the appropriate weights will engender debate over the nature of the current demand curve. The firm will have an incentive to argue that demand curves are very inelastic. This will minimize the penalty associated with raising prices and it will maximize the credit from lowering prices.

Further analysis by Neu (1993) strengthens the conclusions of Brennan. Neu shows that when different products are subject to different rates of growth in demand, the price structure will not converge to Ramsey prices and welfare may be reduced from that under a regime of FDC pricing. Compared to Ramsey relative prices, the firm can earn higher profits by raising the relative prices of goods with higher growth rates and lowering the relative prices of goods with lower growth rates.

Sappington and Sibley (1992) consider the incentives for a firm with pricing flexibility to engage in strategic non-linear pricing. The analysis considers a single-product firm. The firm uses a two-part tariff with a fixed access fee (E_t) and a per-unit commodity charge (p_t).[90] The firm is subject to the equivalent of a Laspeyres price index. The constraint is:

$$p_t + \frac{E_t}{q_{t-1}} \leq AR\,, \tag{5.4.3}$$

90 The consumers' total bill equals $E_t + p_t q_t$, where q_t is the amount purchased. This two-part tariff yields non-linear prices since the average price paid per unit is not constant.

where AR is the constraint on the average revenue of the firm and q_{t-1} is the sales of the firm in the previous period. This constraint can be rewritten as:

$$E_t \leq [AR - p_t] q_{t-1}. \tag{5.4.4}$$

Sappington and Sibley show that the firm has an incentive to lower the commodity charge today in order to charge a higher access fee in the future. By lowering the commodity rate today, quantity expands and this relaxes the constraint on the access charge in the future, allowing the firm to charge a higher access fee. They show that this can lead to a substantial decrease in consumer surplus which is greater than the increase in the profits of the firm. The importance of this result is the general observation that firms will have an incentive to strategically manipulate intertemporal linkages.

If this issue were put aside, it would seem that allowing individual pipelines to choose the relative rate structure will result in a rate structure closer to Ramsey prices than is the case under fully distributed cost pricing. A move towards an efficient pricing structure increases both profits and consumer surplus since the quantity distortions related to raising prices above marginal cost, while meeting the revenue requirement of the firm, are reduced. Therefore, these prices have one attribute of efficient pipeline tolls.

The reconciliation of these prices with the flexibility required to reflect changing demand and capacity may pose a problem. However, the solution may be to allow pipelines the flexibility to set firm transportation rates subject to a non-increasing Laspeyres price index. This would mean that the pipeline has the flexibility to set the structure of average rates. Interruptible rates might then be used to provide some of the required flexibility. However, using a non-increasing Laspeyres price index to set relative rates, while attractive from an efficiency perspective, may run afoul of equity concerns. The actual implementation of such a scheme would require considerable further study.

Optional Tariffs. One method that has been suggested as a way to improve the allocative efficiency of FDC pricing is the introduction of optional tariffs. The firm is constrained to offer the pricing schedule approved by the regulator. However, it is also allowed to offer other pricing options or schedules. Acceptance of one of the schedules by a customer is mutually beneficial since it is a voluntary exchange. The firm would only offer alternative pricing structures if they increased its profits and consumers would only accept an alternative if their surplus was raised by doing so.[91]

---

91 See Brown and Sibley (1986) for further discussion.

As Brown, Einhorn, and Vogelsang (1991) note, the ability to make these kinds of offers gives, in the short run at the very least, an incentive to reduce costs and provide new services. How these optional tariffs might affect the required tariffs has not received much attention under conditions of uncertainty. Moreover, without some re-setting of the mandatory rates, reductions in costs will not benefit consumers.

The use of optional tariffs is a means to allow the firm to effectively deregulate relative prices. Further, relative to other approaches to introducing greater price flexibility, optional tariffs give rise to fewer concerns about equity since the firm must continue to offer regulator-approved prices derived according to traditional procedures. This gives the regulator the ability to protect consumers with inelastic demands.

Priority Pricing.[92] Under priority pricing, consumers are offered a menu of choices relating tolls to reliability of service. *Priority Service* contracts specify the order of priority for each consumer class and the toll charged reflects the order of service. The higher the priority chosen (that is, the less likely the customer is to be interrupted), the higher the demand or access charge per unit. In the simplest design, each consumer selects either interruptible or non-interruptible service. In situations of excess demand, non-interruptible customers are only interrupted after all of the interruptible consumers have been excluded from service.

More complicated designs allow each customer to specify the percentage of their demand which is interruptible and the percentage which is firm or non-interruptible or they permit a number of different priority classes (limited interruptible would be an example). Either a premium can be charged for service which is more reliable than a base level, a discount can be offered for service which is less reliable than a reference level, or some combination. Under these circumstances the demand charge can be interpreted as the price of reliability. In effect, there are two markets, one for the transmission of gas with price equal to the energy charge, and one for each level of reliability with the price given by the appropriate demand charge.

Priority service has a number of potential advantages over contracts where prices and the reliability of service are uniform across customers. The first is that interruptible service or priority service contracts allow consumers to trade off price levels and service reliability levels. Service can be effectively differentiated on the basis of reliability. The increase in product differentiation brings about a closer match between the

---

92 See Wilson (1993) for a full discussion of priority pricing.

quality of service available and the preferences of customers, with resulting efficiency gains. Further efficiency gains are probable if consumers can adapt to their selected level of reliability by changing production technology.

Second, offering a menu of different prices for different levels of reliability can increase the efficiency of capacity allocation. Priority pricing institutes a capacity allocation scheme which, when capacity is over-subscribed, ensures that those who place the highest value on capacity are assured of service. That is, by purchasing the higher-reliability service, they will in effect be outbidding those who put a lower value on the capacity. Wilson (1989, 22) suggests that the number of priority service classes required to realize most of the gains from efficient rationing is quite small and manageable.

A third benefit is that, since offering a priority service menu allows customers to *self-select,* information about the willingness of customers to pay for reliability is revealed. By observing the choices that consumers make from the menu, information about the magnitude of rationing costs they would bear for different levels of reliability can be determined. It can be shown that the premium that a higher service class pays over the amount paid for a lower class of service equals the expected loss in profits of the lower service class from service curtailment (Wilson 1989).

Finally, these contracts allow the pipeline to displace low-priority loads rather then expanding capacity in order to provide reliable service to consumers who place a high value on reliability. As Wilson points out, "... like reserve capacity, contracts for low-priority service provide an inventory of 'supply' to meet shortfalls and thereby protect the higher reliability expected from high-priority contracts." (Wilson 1989, 3).

In an application, TCPL proposed, and the NEB agreed, to implement (NEB 1992b, 18–19) a "more flexible, market-sensitive toll design" for interruptible service in order to "promote a more efficient use of the TCPL system." Two features of this system should be noted. The first is that TCPL would offer multiple service and toll tiers. Higher tolls would correspond to higher-priority service. Second, in order to make the system more flexible, available interruptible capacity would be split into two pools. One pool would be available on a monthly basis, the second on a weekly basis.

A serious impediment to introducing priority service arises when reaching the end-market requires that the gas be transported by more than one pipeline. In these circumstances, the practicality of priority service is jeopardized unless there is sufficient coordination across pipelines. This may be intractable if the pipelines are in different jurisdictions.

Anonymous Mechanisms. This refers to schemes which have been designed to implement efficient pricing for *single product* firms.[93] The initial scheme was proposed by Loeb and Magat (1979). Their proposal involved the regulator paying a subsidy to the firm equal to the total surplus created. In order to get the largest subsidy, a firm would price at marginal cost. Thus, this scheme allows the first-best allocation to be achieved since it assigns title to total surplus to the firm. For the scheme to work, the regulator must know the demand curve in order to determine social surplus. This is, of course, problematic. Moreover, regulators seldom have the ability to pay subsidies. The equity objections to this scheme are obvious.

Further refinements by Finsinger and Vogelsang (1985), Sappington and Sibley (1988), and Sibley (1989) are designed to implement the optimal two-part tariff. These authors show that first-best efficiency can be obtained by paying the firm a subsidy equal to the change in total surplus created by any price change. If the demand for usage is completely inelastic, the required subsidy can be financed by the access component of a two-part tariff. Typically, the strong efficiency results are based on the unrealistic assumptions that the firm produces only a single product, that consumers are homogeneous, and that the environment is stationary. Moreover, while Sibley (1989) has suggested that investment decisions will be close to efficient, this literature has not considered the case when investment is lumpy.

At this stage, the properties of these schemes in real-world situations are not well understood and, as a consequence, their immediate relevance is questionable. One of the important points made in this literature and the literature on optional tariffs is that there are efficiency gains from non-linear pricing.

93 Sibley (1989) is an excellent introduction.

6 Evaluation of Incentive Regimes

The objective in this section is to provide a general evaluation of the main incentive alternatives to traditional Cost of Service (COS) regulation that were presented in Section 5. The evaluation criteria are those outlined in Section 3.4 and used in the evaluation of traditional Cost of Service regulation (see Section 4).

Section 6.1 provides a summary of experience with common types of incentive regulation. The case studies presented involve various forms of price cap regulation, formula pricing, banded rates of return, capital cost incentives, and regulatory streamlining. An evaluation of incentive schemes using price level regulation is provided in Section 6.2. Incentive schemes involving profit level regulation are discussed in Section 6.3. The final section deals with price versus profit regulation in practice.

6.1 EXPERIENCE WITH INCENTIVE REGULATION: CASE STUDIES

Much can be learned about the various regulatory approaches that might be applied in the regulation of natural gas pipelines from an examination of the various schemes implemented in other jurisdictions or in the regulation of other industries. In addition to highlighting the range of options, particularly for streamlining and incentive alternatives, such an examination can be helpful in identifying problems and pitfalls that should be addressed if these types of schemes are considered in the regulation of major Canadian gas transmission systems.

The case studies outlined below are presented with these objectives in mind. The cases include incentive regulation mandated by the Federal Communications Commission; approaches used by the individual U.S. states in the regulation of telecommunications; incentive regulation of U.K. telecommunications; incentive regulation of British Gas; incentive regulation applied to the Tuscon Electric Power Company;

regulatory approaches for the U.S. oil pipeline industry; the regulation of Canadian oil pipelines, including the incentive scheme recently adopted for Interprovincial Pipe Line Inc.; and the regulation of construction of U.S. gas pipelines.

FCC Mandated Incentive Regulation. In the late 1980s, the Federal Communications Commission (FCC) began studying the applicability of alternative regulatory regimes, such as price cap regulation, to the telecommunications companies within its jurisdiction. In 1989, it applied a price cap (PC) scheme to American Telephone and Telegraph Company (AT&T) and, in 1990, it adopted a PC approach with profit sharing (PC-PS) for the regulation of the eight largest interstate local exchange carriers (LECs).

Regulation of AT&T. Prior to 1960, the FCC maintained a policy of 'continuing surveillance' over AT&T rather than employing any specific system of regulation (Brenner 1992, 105). In the mid 1960s, rate-based Rate of Return regulation was adopted. This was employed until 1989 when the FCC switched to the use of price caps to regulate AT&T. This price cap approach imposed a ceiling on prices based on change in the Gross National Product Price Index (GNP-PI) less 3 percent. The 3 percent subtracted from the change in GNP-PI was composed of a 2.5 percent productivity offset and a 0.5 percent Consumer Productivity Dividend (CPD). The 2.5 percent productivity offset was greater than the 2.3 percent that had been estimated for post-divestiture AT&T. The Consumer Productivity Dividend (CPD) was designed to help return some of the benefits of productivity increases to the consumer.

Service Baskets. This price cap plan involved the creation of three 'service baskets.' Basket 1 included the various services to residential customers. Basket 2 grouped 800-number services and basket 3 contained all business services excluding 800 numbers. Caps were set on each basket in order to prevent cross-subsidization from occurring between services offered in different baskets (that is, to prevent cross-subsidization of the competitive business services by residential services where there was less competition). Price caps were not set on each individual service because of the immense number of services (AT&T defined 68,000 individual services) and because caps on baskets would allow greater pricing flexibility. In addition, upper and lower band limitations were placed on service categories, generally limiting the maximum price change of any single service to 5 percent.

The distinct service baskets instituted by the FCC were well suited to serve as transitional stages to a simpler regulatory framework because reduced regulation could be applied in stages starting with the services

		NO SHARING	50% SHARING	100% SHARING
PRODUCTIVITY OFFSET	3.3% (low)	<12.25%	12.25% - 16.25%	>16.25%
	4.3% (high)	<13.25%	13.25% - 17.25%	>17.25%

FIG. 6.1 RATE OF RETURN AND PROFIT SHARING FOR LECs

subject to the greatest amount of competition. In 1991, the price caps on Basket 3 were replaced by 'streamlined regulation' which permitted AT&T to change its rates on 14 days' notice and allowed AT&T to offer single-customer service offerings. Following the introduction of 800-number portability (among AT&T, MCI and Sprint), Basket 2 services also underwent the transition from price-cap regulation to 'streamlined regulation' in May of 1993 (Wald, 1993).

Review of the Price Cap Plan. In its review of the price cap plan, the FCC found that it was successful in achieving the goals of "... reasonable rates, effective incentives for efficiency and innovation, and reducing regulatory burdens." (FCC 1992, 5324). Rates in all three baskets remained below the limits imposed by the price caps, with residential rates declining by 4.4 percent. Consumer benefits were estimated at $1.2 billion dollars over the initial three years. The FCC concluded that no changes to the price cap, including the *X* factor, were necessary, given the decline in prices and relatively low earnings by AT&T over the period 1989–92 (Neri and Bernard 1994).

The FCC also seemed satisfied with the level of expansion of AT&T and the introduction of new services within the price cap plan. With regard to service quality and network reliability, the Commission noted that quality (as measured by an index of the frequency a call is blocked) had declined slightly since the introduction of price caps. The Commission also noted that AT&T's share in the interstate market had declined since the introduction of price caps.

Local Exchange Carriers (LECs). In 1990, the FCC adopted mandatory price caps for the eight largest interstate LECs (the seven 'Bells' plus GTE) to take effect on January 1, 1991 (FCC 1990). Other interstate LECs were also given the option of moving to a price cap system of regulation. The price cap system devised for the LECs included a sliding-scale component as summarized in Figure 6.1.

Under the plan, the unitary or base rate of return is 11.25 percent. An LEC can attain a 14.25 percent effective rate of return if it chooses the low productivity offset and 15.25 percent if the high offset is chosen. There is also a low-end adjustment mechanism whereby, if earnings fall below 10.25 percent, there will be an automatic upward adjustment to the price-cap (a reduction in the productivity offset or the CPD). Services of the LECs are grouped into four baskets: (i) common line, (ii) traffic sensitive, (iii) special access, and (iv) inter-exchange services. The inter-exchange services basket has a 3 percent productivity offset applied to it. The July 1, 1990 rates were used as a basis for the first price cap filing.

State-Level Incentive Regulation in the U.S. Telecommunications Industry. Numerous states have implemented incentive regulation schemes of one sort or another for telecommunications services under their jurisdiction. To date, more than half of the states have made use of regulatory structures other than traditional COS regulation. The following summarizes the experiences of some of the states.

California. In 1989 the California Public Utilities Commission (CPUC) ruled that incentive regulation should replace traditional COS regulation for Pacific Bell and GTE California. The pricing flexibility of the incentive regulation scheme would extend only to competitive services; basic monopoly services would still have close regulatory oversight. Under the price cap, a GNP price inflation index would be used and the productivity offset would be 4.5 percent. One-half of the earnings above a benchmark return would be refunded to utility customers, along with any earnings above a cap set at 5 percent above the market-based rate of return. If the utilities earnings fell 3.25 percentage points below the market-based rate of return two years in a row, the utility could request a review of the inflation or productivity factors. (*Public Utilities Fortnightly*, 1989a)

When the price-cap scheme was adopted on October 12, 1989, the benchmark rate of return was set at 13 percent (1.5 percent above the expected market-based rate of return for 1990), while the maximum rate that allowed sharing was set at 16.5 percent (*Public Utilities Fortnightly,* 1989b). Monopoly services were to be unbundled and made available on a non-discriminatory basis to potential competitors. The services of the LECs were divided into three categories, allowing for a distinction among basic monopoly services, partially competitive services, and fully competitive services (Norris, 1990). In 1992, GTE California and Pacific Bell requested that certain non-pension 'post-retirement' benefits be considered as recoverable costs outside of the price cap. In 1993, the Coalition for Ratepayer Equity (CARE) demanded that the CPUC lower Pacific Bell's rate of return from 11.5 percent to

9.75 percent to reflect the lower cost of money. CARE also requested that the productivity offset be increased above 4.5 percent (O'Shea, 1993).

Illinois. Illinois Bell proposed an alternative regulatory plan with a range of acceptable values for return on equity (ROE) rather than a single target ROE. Under the plan, rates would be set to earn 14 percent, with a baseline of 13 percent. Earnings in excess of the target would accrue to Bell, but earnings exceeding 15 percent would be split between Bell and ratepayers through a yearly refund. The company would not be able to file for a rate increase if earnings remained above the 13 percent threshold. The State Commission initially rejected Illinois Bell's proposal and instead proposed that an earnings-sharing mechanism would be instituted whenever earnings were in excess of the target ROE of 12.76 percent, with a cap of 13.79 percent. This proposal was rejected by the courts because it incorporated retroactive refunds (*Public Utilities Fortnightly,* 1990d).

The sharing structure that Illinois Bell was allowed to implement had Bell retaining 60 percent of additional earnings, reflecting an ROE of between 12.76 percent and 14 percent. Between 14 percent and 15 percent, 30 percent of any additional earnings may be retained by Illinois Bell. Above 15 percent, all additional earnings are 'flowed back' to ratepayers (*Public Utilities Fortnightly,* 1991a). In 1992, the deregulation plan was terminated by the Illinois Commerce Commission and the rates resulting from a 1989 rate order were reaffirmed (*Public Utilities Fortnightly,* 1992).

Michigan. From 1980 through to 1982, incentive regulation involving a price cap was tested for telecommunications in Michigan. The mechanism adjusted intrastate rates across-the-board according to 0.9 *x* (CPI-U – 4 percent). The inflation index used (CPI-U) was the Consumer Price Index on Urban purchases. This plan was estimated to have reduced total real inputs by 1.3 percent in the short run and by 2.3 percent by 1982. Regulatory burden seems to have been reduced over this period as evidenced by a reduction in the time required for hearings. Earnings performance of Michigan Bell was clouded somewhat by the recession occurring at the time of the price cap trial. The estimated cost savings in 1982 were $40 million. The price cap plan was discontinued in 1984 because of divestiture of AT&T and a drop in the CPI-U to 3.8 percent (Face, 1988).

An interim incentive regulation plan was approved by the Michigan Public Service Commission in 1990. Under this plan, a benchmark ROE is set at 13.25 percent. Between 13.25 percent and 14.45 percent, the

LEC retains one-quarter of earnings, one-quarter is shared with ratepayers, and one-half is dedicated to construction programs. Between 14.25 percent and 17.25 percent, one-half is allocated to the LEC and one-quarter is allocated to both the ratepayers and construction programs. In excess of 17.25 percent, one-quarter of earnings is retained by the LEC and ratepayers are entitled to the remaining 75 percent (*Public Utilities Fortnightly*, 1991a). Depreciation regulation and fully allocated embedded costs were maintained as the basis for pricing noncompetitive services (*Public Utilities Fortnightly*, 1990e).

North Dakota. The North Dakota Public Service Commission implemented a price cap called the '1990 Essential Telecommunications Price Factor.' Under this scheme, the price of 'essential services' is regulated by the annual change in general input costs less 50 percent of the annual change in telecommunications productivity (*Public Utilities Fortnightly*, 1990b).

Washington. In 1988, the Washington Utilities and Transportation Commission issued guidelines for proposals concerning incentive regulation (*Public Utilities Fortnightly,* 1988) and, in 1990, an incentive regulation scheme for US West was approved. Under the plan, a five-year cap was placed on rates for residence and business exchange lines, complex lines, the carrier common line charge and access charges. In addition, the plan included a range for the authorized rate of return of 9.25 percent to 11 percent and a formula to determine the sharing of earnings.

Also in 1990, the Commission rejected a proposal by GTE Northwest to institute incentive regulation. Under the GTE proposal, the services of the carrier would be divided into two categories. The first category would be comprised of basic services, the prices of which would be adjusted annually under an inflation index. Non-basic services, forming the second category, would be banded. Earnings sharing would also be instituted, along with a benchmark rate of return (*Public Utilities Fortnightly,* 1990c).

Wisconsin. Wisconsin Bell operated under a trial alternative regulation program from August 1987 to July 1989. Under the program there was a two-year moratorium on rate increases. The ROE was set at 13.5 percent, with retention by the shareholders of all extra earnings up to 14 percent. Earnings between 14 percent and 15.5 percent were shared equally between shareholders and ratepayers, while all earnings in excess of 15.5 percent were to be allocated to ratepayers. In 1990, the LEC was ordered to refund $28.1 million to ratepayers, to account for excess earnings (*Public Utilities Fortnightly,* 1990a). A new plan that freezes local rates was in effect from 1990 to 1994 (Davis 1993, 163).

Other States. As of 1993, Florida, Idaho, Kentucky, Rhode Island, and West Virginia had renewed incentive regulation schemes. Alabama, Connecticut, District of Columbia, Georgia, Louisiana, Maryland, Mississippi, Missouri, Nevada, New Jersey, New Mexico, New York, Oregon, South Carolina, Tennessee, and Texas all have adopted an incentive regulation scheme (Davis 1993).

Incentive Regulation of British Telecommunications. British Telecom, the dominant telecommunications firm in the U.K., has been regulated by a price cap since privatization in 1984. The Office of Telecommunications (Oftel) set the price cap as RPI-X, with the productivity offset (X) set to 3 percent initially (RPI is the percentage change in the Retail Price Index). The price cap applied to a basket composed of residential and business rental charges and directly dialed local and domestic long-distance calls (Hurst 1992, 14). In addition, charges for domestic rentals were not permitted to increase at a rate exceeding RPI + 2 percent in a given year. No similar maximum rate of increase was placed on local call charges. Overall, the price cap affected 50 percent of British Telecom's revenue.

Weights were assigned to each regulated service in proportion to its contribution to turnover in the previous year (Vickers and Yarrow 1988, 213). Prior to the application of price caps, British Telecom was permitted to rebalance prices and costs to a considerable extent, bringing the initial prices closer to costs and reducing the amount of distortion (Bhattacharyya and Laughunn 1987, 28). This differs from the U.S. experience of applying price caps to the tolls most recently approved by the regulator. As well, British Telecom has been allowed to de-average toll rates to a certain extent, as well as institute price differentials for time of use (peak, standard, and cheap) and for two types of long-distance routes differentiated by density. Oftel, in a review in 1986, found that the price differentials reflected costs (Bhattacharyya and Laughunn 1987, 26).

Regulatory Recontracting. Following a review of price cap regulation in 1989, several changes were made to the price cap formula. The productivity offset was increased to 4.5 percent (from 3 percent) and operator-assisted calls were added to the basket, increasing the proportion of revenues under regulation by 5 percent (Hurst 1992, 15).

In September of 1991, international charges were to be reduced by 10 percent and then included in the price cap formula, which itself was revised by changing the productivity offset to 6.25 percent. The RPI + 2 percent limit on increases in domestic line rentals continued for single-line connections only; multiple lines (businesses) had the individual cap increased to RPI + 5 percent (Hurst 1992, 19). Effective August of 1993, while inflation was running at under 2 percent, the productivity offset was further increased to 7.5 percent. The current pricing formula will be in effect for three years and requires British Telecom to introduce its price changes so as to be equivalent to a single price reduction on November 1 of each year (*Times*, Jan. 7, 1994). Under the current price cap formula it is expected that British Telecom will increase the rental price of lines (as permitted) and reduce the price of calls in order to limit revenue loss.

Incentive Regulation of British Gas. Coinciding with the privatization of British Gas in 1986, there was a switch to a price cap regulatory framework under the authority of the Office of Gas Supply (Ofgas). British Gas is vertically integrated with control running from the supply of natural gas in the North Sea fields all the way to the burner tip. The activities of British Gas of concern to regulators can be divided into three main areas.

Services to Large Industrial Users. Services provided to large industrial (non-tariff) customers were not placed under a price cap following privatization but, rather, continued to be regulated in the pre-privatization fashion. Consumers of greater than 25,000 therms per year entered into individual confidential contracts which British Gas priced carefully according to the willingness to pay and availability of alternatives to the individual customer. Initially British Gas was not required to publish its tariffs for its contract segment of the market. In 1988, the Monopolies and Mergers Commission (MMC) ruled that British Gas must publish a schedule of pricing for non-tariff customers (*Times*, Dec. 21, 1988). In May of 1989, British Gas began operating under its first Contract Gas Pricing Schedule (Heal 1990).

It appears that the Contract Gas Pricing Schedule is set by British Gas and subject to review on a complaints basis. The intent of forcing British Gas to publish such a schedule seems to have had more to do with increasing transparency and reducing price discrimination than with achieving any goals regarding efficiency (Price 1992). Under the first schedule, firm contracts were differentiated according to volume (11 volume bands) and number of premises (9 classes of customers), as well as duration of contract and timing of price changes (3 choices). Interruptible contracts were differentiated according to volume and length of interruption (with 18 different prices quoted (Heal 1990)).

Services to Residential and Small Commercial Users. Services to customers using less than 25,000 therms per year were placed under a price cap on April 1, 1987. The maximum allowable rate of change of prices was given by the Retail Price Index (RPI) less a productivity offset (*X*). Any costs associated with the purchase of gas were allowed to be passed through to the customer at their full value. Initially the productivity offset was set at 2 percent. By April of 1991, domestic tariffs were 14 percent lower relative to the RPI than they were at the time of privatization (*Times*, April 30, 1991). In April 1991, the productivity offset to the service element of British Gas charges was increased to 5 percent and the company was no longer permitted to pass the full increase of the cost of the fuel through. Instead, charges for fuel were restricted to changes one percentage point below any rise in the cost of the gas itself (*Times*, April 30, 1991). This represented an attempt to encourage British Gas to look for cheaper sources of supply. However, in January of 1994, the productivity offset for British Gas was reduced to 4 percent, while those customers subject the price cap were narrowed to those using less than 2500 therms per year (*Times*, January 28, 1994, 22). Beginning in 1996, the company will no longer hold its monopoly in the area of sales of gas to households. In 1996 and 1997 all businesses using fewer than 2,500 therms a year will be free to seek supplies outside of British Gas. Independent suppliers will be able to sign up 5 percent of the company's household customers on a first-come, first-served basis in each year. By 1998 it is anticipated that full competition will be achieved.

Pipeline Capacity. In 1982, under the Oil and Gas Enterprise Act, common carrier status was conferred upon gas pipelines in Britain. However, no third-party carriage was ever carried out prior to privatization in 1986. Despite the fact that most offshore gas production was operated by independent companies (fields held by British Gas amounted to only 5 percent of demand), almost the entire supply of North Sea gas was committed to British Gas until 1990 (Terzic and McKinnon 1988). In October of 1988, the MMC recommended that British Gas be allowed to contract for no more than 90 percent of any new North Sea gas fields. There was consideration of obligating the company to purchase any remaining portion of the 10 percent after a certain amount of time had lapsed, but this was not instituted (Heal 1990). One year later, it was reported that North Sea producers were having difficulties finding customers for the 10 percent that British Gas was not allowed to purchase and gas producers were being forced to bear the costs of the government's attempts at bringing about increased competition in the gas industry (*Times*, Oct. 2, 1989). The rate of return that British Gas was entitled to earn on its pipeline assets was set at 4.5 percent by Ofgas in 1989. The company claimed in late 1992 that it required 6.7 percent in order to maintain and improve its network (*Times*, Feb. 25, 1993). British Gas third-party carriage rates (first published in 1990) make use of two-part tariffs, incorporating constant and distance related factors. The element related to distance depended only on load factor while the constant depended upon quantity and which part of the distribution system was used. Changes to the carriage rates in 1990 involved reductions in the distance-related element (Price 1992).

Incentive Regulation Applied to Tuscon Electric Power Company. The most thorough assessment to date of the effects of price cap regulation is that of Issac (1991). He provides a case history of the problems which developed when price cap regulation was applied to the Tucson Electric Power Company (TEP) in the 1980s.

Price cap regulation was adopted because COS regulation was failing to keep nominal or current dollar rate increases under control. Expansion of the rate base due to the inclusion of a nuclear power generating system had resulted in rate shock in the Phoenix area. Price caps were an attractive alternative to the Arizona regulators since they would then avoid the stress and pressures of rate hearings and it would keep nominal prices under control. TEP agreed to a five-year moratorium on rate increases in 1984.

Isaac's analysis of the history of the price cap applied to TEP finds supporting evidence for a number of problems related to manipulability, commitment, counter-factual rate restart, rate shock, and financial distress due to an errant index problem (these issues are discussed in more detail in Section 6.2). His conclusion is that price cap regulation worked well at the beginning, with consumers protected from rate increases, the firm earning high profits, and the regulator avoiding rate hearings. However, over time pressures began to build and, by the time of the review, the system had collapsed amidst acrimony. The two factors primarily responsible for the collapse of the system were cost increases and the negative effect which the recession had on demand. At the initial rate hearing after the expiration of the price cap, TEP requested a total rate increase of 33 percent over two years. After the expiration of the price cap, there was considerable second-guessing about the behaviour of TEP during the price cap period, the review / rate hearings were "nasty," there were a number of court cases initiated by TEP to appeal the outcome of some of these hearings, and, in the "most bizarre twist," the regulatory body appears to have punished TEP for abiding by the terms of the agreement and not raising rates during this period (Isaac 1991, 194).

Regulatory Approaches for the U.S. Oil Pipeline Industry. The oil pipeline industry differs considerably in structure from its natural gas counterpart. Oil is not a homogenous product and it is shipped in particular batches belonging to a specific owner. Delivery in most pipeline systems is, therefore, not instantaneous. Pipelines are not the only economical means of transporting oil. Transport by water (ship or barge) serves as an alternative, with comparable costs (geography permitting). The entire oil industry is quite highly integrated, with most oil pipelines owned by oil-related companies. In 1976, 95 percent of all crude oil shipments and 78 percent of all refined product shipments were carried in pipelines owned by the twenty largest integrated oil companies (Hansen 1983, 18).

Early Regulation. The Passage of the *Hepburn Act* amendment to the *Interstate Commerce Act* in 1906 brought regulation of interstate oil pipelines under the jurisdiction of the Interstate Commerce Commission (ICC). Under the *Hepburn Act*, interstate pipelines were: (i) made common carriers, (ii) banned from practicing undue discrimination among shippers, and (iii) prohibited from providing rebates to shippers (FERC 1982, 61, 588).

Unlike public utilities, oil pipelines were not subject to review or approval of construction, acquisition, abandonment or sale of facilities, presumably in order to allow market forces to play a more prominent role. COS was used to set rates on a point-to-point basis, making use of a valuation rate base.[94]

In the 1940s, the Department of Justice charged three pipelines for violations of the *Elkins Act*. It was claimed by the Department that dividends paid out to owners who were also shippers constituted rebates and, as such, were prohibited. Resolution came about through a consent decree reached by the parties, stipulating a limitation of dividend payments to 7 percent of valuation. Excess earnings were sterilized in a non-earning account restricted to the retirement of debt or covering the costs of new construction. For 16 years, the 7 percent limit was interpreted to apply to the 'total valuation' rather than strictly to the equity-financed portion of the capital base (FERC 1982). This provided the pipelines with the incentive to forego equity financing in favour of debt financing and increase their return on equity. In 1939, debt accounted for 11.8 percent of total capitalization while in 1976 debt accounted for 77.1 percent of total capitalization (Hansen 1983, 30).

---

94 See Navarro, Peterson, and Stauffer (1981).

Regulation by the ICC was minimal. Complaints were relied upon in order to initiate a rate investigation under the presumption that, without a complaint, the status quo was acceptable to all interested parties. Two rate hearings in the 1940s set the allowed rate of return at 10 percent for refined product transportation and 8 percent for crude oil (nominal rates) (Navarro, Peterson, and Stauffer 1981). In the absence of any rate hearings prior to the 1970s, these rates of return became the traditional criteria used by the ICC (FERC 1985, 61,832). Between 1970 and 1977, 14,500 proposed pipeline tariffs were filed, and of these 99.6 percent went into effect in the form proposed by the pipeline companies (Hansen 1983, 21).

FERC Regulation. Following the formation of the Department of Energy in 1977, responsibility for oil pipeline regulation was transferred to the Federal Energy Regulatory Commission (FERC). The FERC made use of two cases that had been initiated in the ICC years in order to refashion oil pipeline ratemaking methodology. In *Farmers Union I*, the FERC adopted a variation of the old ICC methodology, reasoning that prevalent market forces would inhibit the attainment of actual rate levels. This Opinion was reversed by the D.C. Circuit in *Farmers Union II*, signalling a refusal to allow "presumed market forces" to form the principal regulatory constraint. In 1985, the *Williams Pipe Line* case, initiated 14 years prior, came to a close with the issuing of Opinion 154-B. In this Opinion, a traditional COS approach to regulation was established. A net depreciated trended original cost method (TOC) was incorporated. In adopting an original cost (rather than a replacement cost) approach, the Commission affirmed that

> ... original cost is the best yardstick to compare an oil pipeline to other oil pipelines, to other industrial companies, to other industries, and to the entire American economy in order to approximate the oil pipeline's cost of capital. (FERC 1985, 61,378)

Under the TOC methodology, all new assets were added to the rate base at original cost and trended. Existing assets were required to undergo a one-time adjustment to arrive at an appropriate base to be trended for the future. The rate of return would be calculated on a case-specific basis as opposed to using the generic rates of return that had been employed by the ICC. This allowed specific reference to be made to a particular pipeline's risks and corresponding cost of capital. Embedded debt costs would be used as the rate of return for the debt financed portion of capital, while a rate of return would be set on equity capital. The commission also opted to use the actual capital structure of the pipeline or its parent company rather than a hypothetical structure. The starting rate base for existing assets was calculated as the sum of a pipeline's debt ratio times book net depreciated original cost and the equity ratio times the reproduction cost portion of the valuation rate base, depreciated by the same percentage as the book original cost rate base has been depreciated. The formula serves as a middle ground between valuation and net depreciated original cost and ensures that equity holders do not benefit from the write-up of debt financed assets (FERC 1985, 61,836).

Switch to Price Caps. The *Energy Policy Act* of 1992 directed (in a very loose manner) the FERC to streamline and simplify the ratemaking methodology for oil pipelines. The response by FERC was to issue, in October of 1993, Order 561 which puts price caps in place for oil pipelines effective January 1, 1995. An industry-wide (excluding the Trans-Alaska Pipeline System) cap on rate changes is intended to be simple and yet generally applicable to oil pipeline ratemaking. Under the cap, a ceiling level for a given year will be established based on the change to the producer price index on finished goods, less one percent. That is:

$$New\ Ceiling\ Level = Old\ Ceiling\ Level * \left(\frac{PPI_n}{PPI_{n-1}} - 0.01 \right) \tag{6.1.1}$$

where n denotes the year previous to the year of adjustment.

Rates may be charged up to the ceiling level and there is no limit on the number of times a rate may be changed. The indexing system must be utilized to change rates, although there is provision for COS rate regulation under extenuating circumstances or market-based rate determination where warranted. The initial rates will be established by a cost of service hearing, or alternatively, the pipeline may file an initial rate without COS justification provided that at least one non-affiliated shipper is in agreement with the proposed rate.

Problems with the Price Cap. There has been considerable dissatisfaction on the part of oil-pipeline shippers, producing interests, large industrial users and refiners to the move to price caps on oil pipelines. The choice of inflation index was particularly problematic, with parties generally incapable of reaching agreement on the appropriate choice of index (*Inside F.E.R.C.* 1993a; *Regulatory Times* 1993). There were also complaints by shippers regarding the difficulty of challenging rates under the price cap system (*Inside F.E.R.C.* 1993b). Further, shippers and pipelines both feel that regulation has not really been streamlined and litigation is likely to increase (*Inside F.E.R.C.* 1993c).

Market Based Rates. In 1991, the FERC accepted, with some modifications, a proposal by Buckeye Pipe Line Company (Buckeye) to institute a regulatory mechanism based on a yardstick approach (FERC 1991, 61,084). Under the Buckeye proposal, price changes in markets where Buckeye lacks significant market power are used to set caps for price changes in the markets where it does have market power. The FERC's modification basically limited the number of markets to which the 'market-based rates mechanism' would apply. Under the procedure accepted by the FERC:

(i) Changes in the average price in Buckeye's competitive markets (over a certain threshold volume) serve as a cap on the allowed change in the average price in less competitive markets. In this way, increases or decreases in the average price in competitive markets would be mirrored by corresponding increases or decreases in the average price in less competitive markets.

(ii) Individual price changes in the less competitive markets could deviate from the competitive average by a predetermined amount.

(iii) Price changes over a two-year period may not exceed 15 percent in real terms, and price changes not exceeding the change in the GNP deflator plus two percent would not be subject to suspension or investigation.

Regulation of Canadian Oil Pipelines. Prior to the establishment of the National Energy Board, the regulation of pipelines was the responsibility of the Board of Transport Commissioners which carried out its regulation of the design and construction of pipelines under the guidelines of the Pipelines Act. Under the Board of Transport Commissioners, Interprovincial Pipe Line (IPL), which began operating in 1950, was never officially declared a common carrier. Tariff compliance by the pipeline was entirely voluntary. With the move to establish the National Energy Board (NEB), Bill C-49 was proposed which would provide the NEB with control over tolls and tariffs of all pipelines under federal jurisdiction. During the 1960s, the NEB 'watched over' financial and accounting information of pipelines but allowed them to set their own tariffs. No hearings were held on rates but they generally matched the desire of the Board (NEB 1984, 24). In 1969, TCPL, Westcoast, and IPL were formally brought under NEB regulatory responsibility. However, it was not until 1976 that pipeline tariffs were regulated (Restrictive Trade Practices Commission 1986, 144). At that time, the NEB began using Rate Base / Rate of Return regulation based on historical costs to determine IPL's tariffs.

Streamlining. There are three classes of Toll Adjustment Applications that have been instituted in order to streamline the regulatory process. Class 1 and 2 applications do not normally require formal public hearings and involve changes to tolls resulting from changes in throughput and significant changes to the cost of service. Class 3 applications do require public hearings and involve changes to the authorized rate of return, and calculations of allowances for income taxes, among other things (NEB 1990b, 23).

Banded Rates of Return and Trigger Mechanisms. The NEB identifies 10 Group 1 pipelines in Canada. These are characterized as extensive systems. Of these, five are oil and products pipelines and include the pipelines operated by Cochin Pipe Lines Ltd. (Cochin), Interprovincial Pipe Line Inc. (IPL), Interprovincial Pipe Line (NW) Ltd, Trans-Northern Pipelines Inc. (TNPL), and Trans Mountain Pipe Line Company Ltd. (TMPL). Two of these, Cochin and Interprovincial (NW), are regulated on a complaints basis. The remaining three have a toll trigger mechanism incorporated in their regulation.

TMPL was the first to receive a trigger mechanism. Under the initial system established in 1978, the company was required to file an application for new tolls whenever it had a 5 percent variation in its throughput. This 5 percent variation in throughput was meant to equate to a 1.6 percentage point variation in return on equity (NEB 1992a, 31). Under the throughput variation trigger, whenever the throughput forecast for the current calendar year (measured in cubic metre-kilometres) varied by 5 percent or more from the throughput on which tolls had been based, TMPL was required to submit new tolls to reflect the revised volume (NEB 1985, 31). The throughput trigger was chosen because the company had a small rate base and was subject to large variability in volume. The trigger was later modified to a 4 percent variation in revenue, which in 1992 approximated a 2 percentage point variation in return on equity. In 1988 and 1989, under the 4 percent revenue variation trigger, TMPL's actual returns on equity were 17.97 and 17.64 percent, respectively, against an approved rate of 14 percent (NEB 1990b, 10).

In 1980, IPL was brought under a trigger mechanism along with Cochin. Cochin tolls were later changed to regulation on a 'complaints basis.' In 1985, IPL's trigger mechanism was amended and applied to TNPL as well. The trigger for these two pipelines is activated by changes in the return on equity of the companies. IPL and TNPL are required to file an application for new tolls when it is forecast that their return on equity for the calendar year will exceed the approved return by more than two percentage points (NEB 1990b, 9). The approved return on equity is that return which the Board has set at the most recent toll hearing. Specifically, a comparison is made between the forecast net income for the year, expressed as a percentage of the approved equity component of the updated actual rate base, compared to the return on equity approved by the Board at the most recent toll hearings. In 1992, TMPL was also brought under the trigger involving a 2 percentage point variation in return on equity.

Tolling. Tolls are set on a point-to-point basis ($ per cubic metre), differentiated into five different commodity groups (heavy crude, medium crude, light crude, Refined Petroleum Products (RPPs), and NGLs). A light-crude toll is set and medium and heavy crudes are assigned a surcharge to this toll. Presently, NGLs and RPPs are assigned a credit relative to the light-crude toll. The surcharge or credit for each commodity type is calculated by devising a hypothetical pipeline, dedicated to transport of that commodity, which provides the lowest 20- year average toll. This is then compared to the existing pipeline and the surcharge or credit is calculated accordingly (NEB 1989). Capacity-related capital and fuel and power cost considerations are integrated for the purposes of surcharge design. In 1987, IPL had 51 different point-to-point tolls for light crude (not including tolls on the Lakehead system in the U.S.), originating from eight different sources. Heavy and medium crude shipments were subjected to 15 and 5 percent surcharges respectively. There were four tolls for RPPs and three for NGLs (NEB 1987, 67).

Incentive Regulation. In March 1995, the NEB approved an application by IPL concerning a negotiated toll settlement.[95] This settlement, which was between IPL and CAPP (Canadian Association of Petroleum Producers), embodies an incentive toll methodology.[96] Briefly, it is essentially a restricted price cap with profit sharing.

This agreement covers the period January 1, 1995 to December 31, 1999, and there is an intention that in 1997 the parameters will be negotiated to extend the arrangement beyond 1999. The scheme begins with an agreement on a gross revenue requirement of $377.2 million for

95 See NEB, Re: Application to the National Energy Board by Interprovincial Pipe Line Inc., dated 16 February 1995, for Orders Pursuant to Part IV of the National Energy Board Act, Approving a Negotiated Toll Settlement, 24 March 1995.

96 See IPL, "In the Matter of an Application by InterProvincial Pipe Line Inc. For Orders Under Part IV of the Act Approving a Negotiated Settlement Respecting an Incentive Toll Methodology and Associated Tolls and Tariffs", application to the NEB, February 1995 and IPL, "Incentive Toll Proposal, Principles of Settlement," effective April 1, 1995.

1995. This is made up of a 'starting point' revenue requirement of $346.4 million (this starting point represents the component which is escalated according to overall inflation rates), a forecast allowance for income taxes (federal and provincial corporate income taxes plus the large corporation tax) equal to $24.2 million, and an Allowance Oil Revenue with an estimated value of $6.6 million. This last item is essentially a toll collected in kind (equal to 1 / 20 of one percent of the volume of hydrocarbons tendered by each shipper) and it is intended to provide for the actual oil losses incurred by IPL.

For each subsequent year, tolls are based on a net revenue requirement which is essentially determined by escalating the net revenue requirement for the previous year[97] by the percentage change in the Consumer Price Index (CPI) and taking account of several variance and sharing items. The change in the CPI is restricted in that it is deemed not to exceed 5 percent nor fall below 1 percent.

The variance and sharing items include a correction for any previous differences between the forecast rate of inflation (which is used to set tolls for the test year) and the rate of inflation that materialized (the *Inflation Adjustment*); a correction for any differences between the forecast tax allowance and the actual income tax paid (the *Tax Allowance Variance*); an allowance for any 'non-routine' expenditures[98] (the *Non-Routine Adjustment Variance*); an adjustment for capacity sharing involving a mechanism which allocates the additional net revenue to IPL and the shippers on a 75 percent / 25 percent basis if throughput exceeds some agreed capacity level (the *Capacity Sharing Amount*); a cost reduction benefit sharing which allocates the earnings above various thresholds to IPL and the shippers (the *Cost*

97 The net revenue requirement for the first year (1995) is $370.6 million, made up of the Starting Point and the Forecast Tax Allowance.

98 These include increases in costs resulting from programs requested by or agreed to by shippers related to such things as quality initiatives, new services or extension of new services, or new preventative maintenance measures whose benefits occur after the term of the agreement; decreases in costs because of changes in the services requested by or agreed to by shippers; adjustments where all or any portion of the Montreal extension ceases to be tolled on an integrated basis; the associated costs where IPL is subject to an NEB order, or is affected by precedent established by an NEB order to other pipelines, that results in increased costs to IPL; changes in costs arising from legislation, regulations, orders, etc., that result in changes to safety or environmental requirements, practices or procedures for IPL; additional costs for distinct or new programs to address new or unanticipated failures of the IPL system; and increases to net revenue to correct or reverse any downgrade of IPL below BBB (low) or equivalent by one of the Canadian bond rating agencies. See IPL (1995, 22–23).

Performance Benefit Sharing[99]); an adjustment for any difference between forecast and realized transportation revenue as a result of decreases, below the level of annual capacity upon which tolls are fixed, in the volumes shipped, a change in the throughput mix or a change in the average length of haul[100] (the *Transportation Revenue Variance*); a deferred amount adjustment and an adjustment for that period in 1995 when the Oil Allowance was not collected; and all carrying charges associated with the foregoing items.

As with other price cap schemes, the intent is to break the link between revenue requirements or tolls and the costs of providing the service. In this case, since revenue requirements are driven by changes in the overall consumer price level rather than by changes in actual costs, there exists a stronger incentive for IPL to reduce costs. As such, any significant reductions in costs will result in higher profits, some of which will be kept by IPL and some distributed to shippers through toll adjustments. The *Cost Performance Benefit Sharing* component (described immediately above) sets out the nature of this sharing. The other major incentive element involves the *Capacity Sharing* item which allows IPL to keep 75 percent of the increase in profits associated with exceeding some set capacity level (however, this is not symmetric: IPL is not penalized if throughput falls below this capacity level).

A complete evaluation of this scheme will not be possible until it has been in operation for a number of years. At this point it is only possible to highlight certain features and indicate their likely consequences.

It can be noted that the price cap employed in this case represents a somewhat different approach to price caps than that used in most other cases described previously. First, it is restricted in the sense that the allowed escalations are between 1 and 5 percent. As such, there will be additional gains to IPL if the actual inflation rate averaged less than 1 percent over the term of the agreement and additional gains to the shippers if the actual inflation rate averaged more than 5 percent. Second, the price cap does not incorporate an X factor. It will be recalled that this factor amounts to a reduction from the change in the

99 Under the Cost Performance Benefit Sharing, IPL retains all of the after-tax earnings up to $51.5 million. Any earnings in excess of this amount but less than $58 million are to be shared on a 60/40 basis by IPL and the shippers. Any after-tax earnings in excess of $58 million are to be shared on a 50 / 50 basis. Any sharing with the shippers takes place through a reduction in the revenue requirement and tolls from what they would otherwise be in the subsequent year.

100 The actual adjustment under the Transportation Revenue Variance allocates the surplus or deficiency on a 50/50 basis using a deemed Power Allowance of 50 percent.

CPI in the allowed rate of change in the revenue requirements or tolls. It is also worth noting that, assuming no other variance or sharing adjustments for the moment, this approach means that the revenue requirement should increase significantly faster than actual costs. For example, while operating and maintenance costs might be expected to rise with inflation, this would not be true for other major items like depreciation or capital recovery (which should decline in the absence of significant expansion of the system) and return on capital.

As discussed in Section 5.1, price cap schemes applied to such things as oil and gas transmission should make provisions for substantive changes which are beyond the control of the regulated utility. The mechanisms in the IPL / CAPP scheme incorporate such provisions. For example, it can be noted that any changes related to income taxes or the large corporation tax are completely compensated so that IPL is not put at risk. While it would appear that IPL is at risk for any substantive changes in certain other taxes (in particular, local property taxes), this must be considered in the context of overall increases in the revenue requirement which will likely exceed increases in costs by a significant margin. In addition, the scheme protects IPL from the consequences of forecast errors (with regard to the forecasts of such things as the Consumer Price Index, income taxes and throughput) and 'non-routine' occurrences. The latter amount to a set of off-ramps to allow for adjustments in the event that, for example, the system is expanded, IPL's credit-worthiness is negatively affected or there are other costs related to regulatory or government policy decisions. It is interesting to note that these generally amount to 'resetting' the formula. For example, in the case of a requested or approved expansion, the 'base' revenue requirement is adjusted to incorporate the owning and operating costs associated with the additional facilities.

Another interesting feature is that this scheme would appear to provide some flexibility in adjusting individual tolls so long as the overall price cap is not violated. It might be recalled that the possibility of relative price adjustments for individual services is considered a significant efficiency advantage of applying the price cap to overall revenue requirements rather than to individual tolls. However, as explained in Section 6.2, such adjustments may create problems with regard to regulatory requirements for such things as just and reasonable tolls and no undue discrimination. Under the IPL / CAPP scheme, the incentive toll mechanism is based on the premise that tolls will still be established using toll designs approved by the NEB from time to time. In addition, the NEB retains the right to override any provisions in the negotiated scheme if it believes such is required to protect the public interest.

Regulation of Construction of U.S. Gas Pipelines. In September of 1991, the FERC approved new rules regarding construction for gas pipelines. Construction of pipeline expansion can now be carried out under a revised Kansas test or under an expedited certificate form.

Early Regulation of Construction. Under the original Kansas test, the pipeline had to show sufficient supply and demand before beginning construction. The traditional Kansas Pipe Line Criteria required an adequate showing that (FERC 1991, 30, 224):

(i) The applicant possesses a supply of natural gas adequate to meet those demands which it is reasonable to assume will be made upon them.

(ii) There exist customers who can be reasonably assumed to use the gas.

(iii) The proposed facilities are adequate to render a full and complete public service in the territory proposed to be served.

(iv) The applicant possesses adequate financial resources.

(v) The costs are adequate and reasonable.

(vi) The rates to be charged in the future should reflect costs.

(vii) The anticipated fixed costs must be reasonable.

Determination of public convenience was based on these standards for fifty years.

Revised Methodology. Under the revised Kansas test, there is less stringent proof required on the actual need of the pipeline. Pipelines which receive a certificate under this test are permitted to pass-through prudently incurred construction costs to the ratepayers. Sufficient market demand has been interpreted as long-term (minimum 10 year) contracts for 100 percent of the capacity of the proposed facility. To establish sufficiency of supply it is necessary to describe the accessible production areas and how they connect to the proposed construction. If an existing pipeline meets the Kansas standards it may establish incremental, stand-alone rates for transportation, or it may maintain its existing rates (FERC 1991, 30, 269). See Section 4.3 (*Incremental Versus Rolled-In Tolling*) for the most recent policy concerning incremental versus rolled-in rates.

Expedited Certificates. For pipelines which do not pass the revised Kansas test, or those that do not wish to wait for certification, permission to build is granted simply by filing notice with the FERC. There is no requirement to show adequate supply and demand. Under this form of construction, pipelines are labelled as 'at risk' for recovery of construction costs and may not pass these costs on to ratepayers. Rates under such a system require a 90 percent throughput of volume. It is possible for a pipeline that constructs in such a manner to no longer be considered 'at risk' by showing a legitimate need for the pipeline (evidence of long-term supply contracts).

6.2 EVALUATION OF PRICE LEVEL REGULATION INCENTIVE SCHEMES

The cases outlined in the previous section suggest that price level regulation incentive schemes, usually with some form of profit sharing, have been the most common incentive alternatives to traditional COS regulation. This section provides a general evaluation of the various price level regulation incentive alternatives set out in Section 5.2.

Price Caps with RPI-X. The standard price cap scheme is where maximum prices are determined according to the change in some economy-wide price index, less some productivity growth or *X* factor. For general evaluation purposes, the version assumed is that where a retail price index (RPI), such as the overall Consumer Price Index (CPI) in Canada, is employed as the measure of overall price changes. There are four features of price-cap regulation which are critical to understanding and assessing its performance.

(i) *Incentives for Cost Minimization.* Rather than profit maximizing subject to a rate of return constraint, under price cap regulation the firm maximizes profits subject to an exogenous price ceiling. That is, the link between prices and costs is severed. This provides a firm with incentives to minimize costs and invest in cost reductions. The benefits from cost minimization and successful investment in cost reduction will accrue to the firm as higher profits instead of, as in COS regulation, lower prices for customers. In addition, the firm should have no incentive to misreport its costs or strategically manipulate the allocation of common costs between regulated and unregulated activities.[101]

---

101 For a formal treatment which compares the incentives for cost minimization and cost misallocation under COS regulation and price cap regulation, see Braeutigam

(ii) *Errant Indices.* To the proponents of price cap regulation, one of its most attractive features is that the firm is encouraged to minimize costs. This is most easily seen by considering the incentives of a single product firm subject to a price ceiling in a static environment. However, in the more common situation where demand and costs change over time, the desirability of price level regulation depends on how the cap is adjusted to reflect both changes in costs and demand. Price cap regulation has two mechanisms whereby the price cap can adjust to a changing environment. In the short run or near term, it is adjusted by an exogenous factor such as *RPI–X,* and in the long run the entire mechanism is subject to review.

The price cap requires adjustment in the short run for two reasons. The first is to protect the viability of the firm and the second is to pass some of the benefits from cost reduction on to customers.

In an inflationary environment, the prices of the inputs of a firm will be rising. Without some latitude to pass on this increase in costs through higher prices, the firm will encounter financial difficulties. Consequently, one of the adjustment factors should reflect the costs of the firm. However, it obviously cannot be based on the actual costs of the firm or the purported incentives for cost reduction and operating efficiency will disappear: the link between costs and prices will be inadvertently re-established.

The indexes usually proposed to capture the effect of inflation on the costs of the firm include the Consumer Price Index (CPI), the Producer Price Index (PPI), or the Gross Domestic Product Price Deflator (GOP-PI).[102] All of these indexes suffer from the same defect. That is, none of them will necessarily capture the specific cost changes that a regulated firm faces. All are economy-wide measures and it is highly unlikely that any one of them will accurately track the impact of inflation on the costs of a particular

---

and Panzar (1989). The issues of strategic cost misallocation and inefficient diversification into competitive markets by the regulated firm are less important in the context of pipelines than in other industries like telecommunications. In telecommunications in particular, the presence of strong economies of scope between regulated and competitive markets provide incentives for firms to diversify into competitive markets. Diversification into competitive markets by pipelines under COS regulation has not been a motivating concern for regulatory reform of pipeline transmission in Canada.

102 The GNP price deflator (GNP-PI) is another possibility but has largely been supplanted by the GDP deflator in Canada.

firm. What they will do is tend to cause the real (or constant dollar) prices to customers to remain constant over time or, in the case where a positive *X* value is used, to decline over time.

This tendency to not track the changes in costs facing the firm regulated under PCs gives rise to what Hillman and Braeutigam (1989, 69) term the "Russian roulette" problem of errant indices. Adoption of one of these indices is likely to result in a tracking problem: that is, a divergence between prices the firm receives and its unit costs. This will result in either windfall gains to the regulated firms or their customers and significant pricing inefficiency.

The productivity adjustment factor is the payoff to consumers from the change in regulatory regimes. It is also a critical component of price cap regulation because, when it is set optimally, it provides a means for prices to more closely track costs and to reduce the need for frequent 'truing up' of prices to reflect actual costs. The question naturally arises as to how the expected productivity gain is to be determined. The historical record is unlikely to be of any use since it was based on a different regulatory regime and a different set of technological opportunities. Moreover, since the productivity offset, *X*, will be fixed until the comprehensive review, too high a choice could threaten the viability of the firm. Between comprehensive reviews, the firm is supposed to bear the risk of meeting *X*. The problem of the errant index is potentially compounded by an inappropriately set value for *X*.

The experience of Michigan Bell in the early 1980s provides evidence of the errant index problem and the resulting financial distress. The value of *X* was set too high (at 4 percent) and Michigan Bell failed to earn its allowed rate of return. Its financial distress was alleviated somewhat by the fact that the CPI index used to adjust the price cap overstated the increase in its input costs.[103] The "Russian roulette" problem of the errant index and an incorrect forecast of *X* is further compounded by exogenous and unforeseen changes in either demand or costs.

(iii) *Periodic Reviews.* A component of every price cap scheme is a promise that the entire regime will be evaluated at periodic intervals. For example, as noted in Section 6.1, the focus of the reviews for British Gas and British Telecom have been on adjusting the productivity or *X* factor. In the first review of British Telecom,

103 See Face (1988) for details of the Michigan Bell experience.

Oftel, its regulatory commission, increased *X* from 3 percent to 4.5 percent. There was no detailed explanation provided, though the Director General of Telecommunications did indicate that the rate of return earned by British Telecom was the most important criterion, if not the only one (Beesely and Littlechild 1989). Subsequently, the *X* factor was raised to 6.25 percent in 1991 and in 1993 it was further increased to 7.5 percent. In the U.S., the FCC reviewed the AT&T cap in 1993 to ensure that "consumers will be beneficiaries of all achieved economies." (FCC 1989, paragraph 42) Though the earnings of AT&T had been relatively low, the FCC did not reduce the *X* factor, but maintained it at 3 percent.

In both of these cases, it seems clear that the profits of the firm do affect the determination of the new cap and the *firm should expect that its profitability will influence the regulator's assessment of the price cap.* In such a setting it would appear that the incentives of the firm to cost-minimize, to adjust its input mix optimally, and to adopt new technologies are substantially reduced. The increase in profits from these investments, just as under COS regulation, is passed on to customers in the form of lower prices, either by an adjustment to the price level or a higher *X*. If the frequency of reviews is very similar to the frequency of rate hearings, any cost inefficiency associated with COS regulation is also likely to be associated with price cap regulation.

Even if regulators were to state that they would not consider profit data, it is hard to see how they could credibly commit to ignore the profitability of the regulated firm. Regulated firms are well aware of the incentives for future regulators to act opportunistically. Indeed Pierce (1990, 693) goes so far as to suggest that there is a real risk under price cap regulation that profit data will be used only when it leads to lower prices and that regulators will not increase the cap when profits decline.

Recent experience in Britain would seem to suggest that, even under regulation by Mr. Littlechild, the architect of price cap regulation, there is a strong tendency to re-set the *X* factor based on rates of return earned by the utility and on public perception. This, combined with the frequency of such adjustments, raises the issue of whether, in practice, price cap regulation has a tendency (even where the regulators are appointed and considered to be independent of government) to revert to rate of return regulation.[104]

---

104 For a recent assessment of the British experience, see The Economist (1995, 74).

Riordan and Cabral (1989) provide theoretical support for the notion that periodic reviews undermine the incentives of the firm to invest in cost reduction. If the firm forecasts that the price cap will be too low and, therefore, unprofitable, even with the socially beneficial new technology, it has an incentive to not invest in the new technology. Instead, it will request a rate hearing when the price cap begins to bind, typically on the basis that its financial viability will be threatened without an upward adjustment in the cap. This example also provides an illustration of the incentive for a firm subject to price cap regulation to manipulate its costs and of the incentive for a regulator to second-guess the decisions of the firm.

(iv) *Price Flexibility.* In some applications it is not individual prices to which caps or ceilings are applied. Instead a cap is applied to a basket of goods / services: the cap applies to a weighted average (the weights are usually the quantities sold in the previous period) of the prices of the goods / services which comprise the basket. This provides the firm with an opportunity to re-balance its rates. This means that the firm will typically find it profitable to raise the prices of goods / services whose demand is inelastic or less price-sensitive and lower the prices of goods / services whose demand is elastic or price-sensitive. This increases profits. Raising the price of the inelastic group increases total revenues since the quantity demanded decreases relatively little; lowering the price for elastic customers will also increase revenues since the quantity they demand increases substantially. In order to take advantage of this opportunity, the firm has an incentive to differentiate its services. However, regulators have often limited the ability of firms to engage in this kind of price discrimination. They do this by carefully constructing the basket of goods / service and by placing limits on the amount by which the price of any one good / service can rise.

Allocative or Pricing Efficiency. It will be recalled that allocative efficiency requires that prices reflect incremental costs. Using historic cost accounting to set the initial prices means that these prices are unlikely to reflect economic or opportunity costs. Furthermore, even if initial prices are set optimally to reflect economic costs, the errant indices problem will result in prices which diverge from economic cost and thus provide inappropriate market signals. A significant concern is the theoretical and practical reconciliation of price caps with two-part tariffs for a multiproduct firm.[105]

Efficient Rationing and Product Selection. Under a price cap scheme where the firm can set relative prices and introduce new profit-maximizing services, the firm will find it can increase profits by practicing second-degree price discrimination. For example, in the context of pipeline service, a firm could do this by introducing priority pricing.

There are two main quality-related concerns associated with price cap regulation. First, a direct consequence of the incentives for cost reduction that a firm faces under price cap regulation is the lack of incentive it has to maintain the quality of service. Reductions in service quality may reduce costs and thereby increase profits. Quality control has been a major concern of Oftel, the regulator of British Telecom. The extent of the problem is illustrated by the recent agreement between Oftel and British Telecom under which British Telecom agreed to accept contractual liability for repair and provision of new services. *Per diem* charges would accrue for service problems of individual customers that take more than two days to fix (Brown, Einhorn, and Vogelsang 1989, 88).

The quality-of-service issue is likely to be even more contentious in the case of a gas pipeline. This is because it will be very difficult to establish an acceptable and meaningful measure of quality of service. The main element of quality of service will be reliability or the absence of unplanned outages. But, since such outages are largely random, their frequency or magnitude in any given period may have little to do with whether or not the firm was increasing profits at the expense of maintaining service quality. In fact, it would take a fairly long period of reliability observations (so that the outages unrelated to maintenance expenditures would 'average out' to zero) in order to detect any systematic changes by the firm with regard to service reliability.

The second quality-related concern has been that expressed about the extent of service coverage. In particular, services which do not cover their incremental costs are prime targets for elimination (Hillman and Braeutigam 1989, 68).

A similar problem will result if the X factor applies to the firm as a whole, instead of having different values for different products. This provides the firm with an incentive to over-produce goods / services where the rate of productivity growth in their provision exceeds X and

105 See Liston (1993) for details.

under-produce or eliminate those products where the rate of productivity growth is less than *X*.

Cost Efficiency and Incentives for Cost Reductions. The incentives under price caps for cost minimization and investments in cost reduction depend on the ability of the regulator to commit to not use profit data when the price caps are reviewed. If such a commitment is not possible, which is most likely to be the case, price cap regulation may well be limited to institutionalizing regulatory lag. As noted earlier, this lag does provide an incentive for cost efficiency but it would already exist in most cases where COS regulation is applied on a test-year basis.

Cost Efficiency vs Allocative Efficiency. Schmalensee (1989c) considers the robustness of various regulatory constraints to the introduction of uncertainty over future costs and asymmetric information about managerial effort to reduce cost.[106] His analysis explicitly compares the total surplus generated under both cost-plus regulatory regimes (COS regulation) and price cap regulation (no cost pass-through).[107] In essence, he compares the social value of the greater incentives for cost efficiency under price-cap regulation to the social value of prices which reflect costs under COS regulation.

Under COS regulation, the regulator commits to set prices in line with observed unit costs. Under price cap regulation, the regulator commits to fixed prices. The regulator commits to the regulatory regime before the firm selects its effort level. After the firm selects its effort level, realized costs, which depend on both the effort level and a random shock (to reflect factors beyond the control of management), are revealed to both the firm and the regulator. Based on realized costs, prices are determined by the rules of the prevailing regulatory regime.

COS regulation does not involve any prior choice of parameters by the regulator, but price cap regulation does. In the price cap scheme, the regulator must determine the price cap before realized cost is observed. In doing so, it must set the price cap high enough to ensure that the firm breaks even. Schmalensee considers two different constraints on profits. The first is that expected return equals that allowed by the regulator; on average the firm earns its allowed rate of return. The second is that

106 The asymmetric information in Schmalensee's model concerns the cost-reducing effort of management. Realized costs depend on a random factor and the effort level of management. A regulator, who only observes realized cost, cannot infer what the effort level of management was. Management, of course, knows its effort level.

107 Schmalensee considers the problem with two different objectives. The first is to maximize consumer surplus and the second is to maximize total surplus (the sum of consumer surplus and profits).

for any actual cost realization, the firm earns its allowed rate of return. This second constraint involves setting the price cap so that even if costs are very high, the firm is assured of earning its allowed rate of return. Under the first constraint, the price cap is set so that the firm earns its allowed rate of return only on average. Hence, if costs are very high it may not earn its allowed rate of return.

Price cap regulation provides a greater incentive for the firm to reduce costs; that is, it gives managers an incentive to exert greater effort. However, holding price constant to provide incentives for cost reduction prevents efficient pricing. Prices cannot respond to actual costs and, as the degree of uncertainty increases, the greater is this social cost. This inefficiency is compounded when the firm cannot be compelled to serve. In this case, the initial price cap must be set to ensure that prices are high enough to allow the firm to earn its allowed rate of return. The higher the price cap, the greater the wedge between realized costs and prices *ex post*. The greater this gap, the greater the lost surplus from not having prices track costs.

Schmalensee concludes that price caps have been "over sold" relative to cost-based regulation. He finds that some cost pass-through is always optimal: price caps are never optimal if there is some uncertainty about costs. Moreover, if the objective of the regulator is to maximize the surplus of consumers subject to the firm earning a normal return for all realized costs, it is more likely that the preferred regulatory constraint will involve significant cost pass-through.

This conclusion is reinforced by the work of Riordan and Cabral (1989). They compare the incentives for management to reduce costs under COS regulation when there is a regulatory lag with those under price cap regulation. Though price cap regulation provides stronger incentives for cost reduction, they show that prices may be lower under COS regulation with a regulatory lag than under price caps, since all reductions in cost under COS regulation are passed on to customers.

Fairness and Equity. Price cap regulation clearly violates the first elements of this criterion, especially if relative price flexibility is allowed and this results in price discrimination. It will be recalled that two common elements of the fairness criterion are that prices for each service should be based on the costs of providing it and that there should not be price discrimination (or significantly different prices for equivalent service).

Moreover, a consequence of the errant indices problem is that when the price ceiling is reviewed it may be the case that prices have not increased as rapidly as costs. In these circumstances, there may be a politically unacceptable *rate shock* if prices rise dramatically upon review. Regulators may prefer small annual increases over the duration of the price ceiling rather than a discrete, equivalent adjustment upon review.

Whether cost reductions actually benefit consumers depends on the value of *X*. The errant index problem will create 'windfall' gains and losses for either the regulated firm or its customers. These windfalls will be due to failure of the price cap to adjust effectively to unexpected changes in costs and demands. Legitimate concerns regarding the fairness of this process will likely be raised.

Economic Viability. Price cap regulation may not measure up very well under either component of this criterion (that the regulated firm's viability is protected and that the regulatory regime is sustainable). The errant index problem, in particular, means that the firm is subject to significant risk and the concerns over the equity problems created by the errant index problem may generate continual demands for change.

The viability of the firm will also depend on whether the accumulated depreciation at the time of the switch to a price cap accurately measures economic depreciation. The switch to price cap regulation locks in the accumulated depreciation, opening up the possibility that it was either too large or too small. If it has a history of being too small, then the imposition of maximum prices which do not allow the firm to make up its depreciation reserve will threaten the financial viability of the firm (Hillman and Braeutigam 1989, 46).

The question of the allocation of risk and the viability of the firm has important cost implications. Increases in risk will likely result in an increase in the cost of capital. This will show up as a decrease in debt / equity ratios and / or an increase in required rates of return. Given the capital-intensive nature of pipelines, an increase in the cost of capital will significantly increase the cost of service. If financial markets dictate a significant increase in the risk premium for a regulated firm, it seems highly improbable that the potential cost savings from incentive regulation would adequately offset the effects on tolls of the increase in capital costs.

Regulatory Burden. There is a presumption that price cap regulation will entail lower administrative and regulatory costs. Between reviews, the price cap is adjusted automatically by a formula, thus effectively eliminating the need for a rate hearing and the attendant costs.

However, whether in fact there will be fewer formal regulatory hearings in a case such as price caps applied to pipelines will depend on the number of capacity expansions. The use of historic costs will in all likelihood mean that significant additions to the rate base will change average (historic) cost and require a change in the revenue requirement. Unless prices are adjusted to reflect the higher costs arising because of significant expansions, the firm is unlikely to provide the expanded service that might be requested based on market considerations.

It is also useful to note that experience, such as that from the use of price cap regulation for British utilities (see Section 6.1), might suggest that the reduction in regulatory burden may be considerably less than expected. In addition to the frequent hearings to adjust the allowed escalation of prices (primarily through changes to the X factor), the administration of a myriad of deferral accounts requires considerable effort by both the regulator and the regulated firm.

Implementation. Price cap regulation would appear to have the virtues of simplicity and practicality. However, consideration of all the different parameters under the control of the regulator belies the simplicity of the approach. Figure 6.2, from Veljanovski (1993), shows the increasing complexity of actual price cap regimes in Britain. As indicated, the simple mechanism in theory (RPI-X) has, in practice, been subject to many complicating modifications. These include the addition of a variety of correction factors such as an investment factor, a gas cost index, an energy efficiency factor, a factor to take account of additional security costs and a cost pass-through factor. In many instances, these represent little more than ad hoc adjustments to make the outcomes of the price cap more acceptable. It is evident that such regimes are subject to growing complexity.

The fact that under price caps the regulator supposedly does not monitor costs and profits significantly reduces the administrative costs of regulating. Failure to regulate profits, however, may result in a lack of acceptance and a loss of legitimacy. In any case, as indicated by many of the case studies presented in Section 6.1, it would appear that, in practice, the regulator will usually continue to monitor profits and costs.

The lack of regulatory involvement in major decisions, like capacity expansions, can call into question the legitimacy of the decisions of the firm. If costs are linked between one time period and another (due to durable sunk investments), firms will have an incentive to shift costs such that they are included in cost-based prices when prices are revised. To counter this behaviour, regulators may have to determine what the costs would have been had the firm not had an incentive to manipulate its costs (Issac (1991) refers to this as counter-factual rate restart). The alternative, as Jones (1991, 2) notes, may be that regulators become even more involved in the day-to-day operations of the firm.

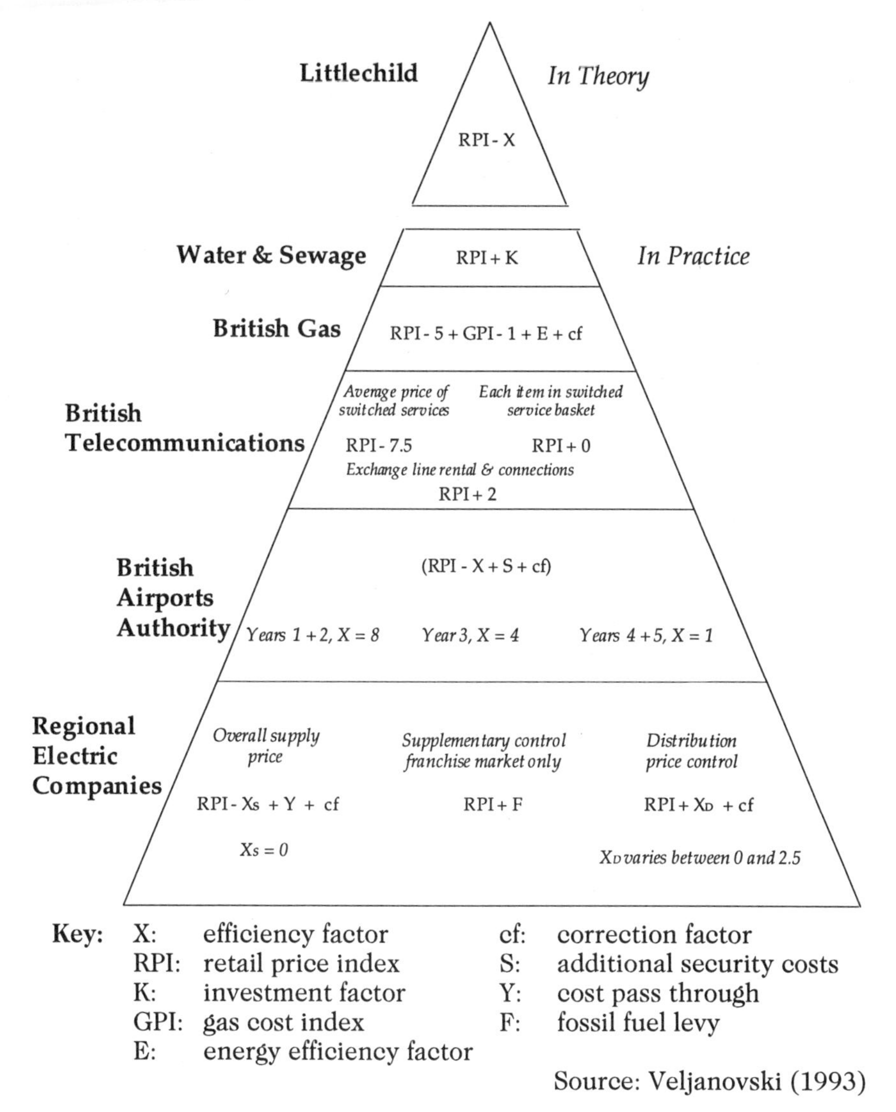

FIG. 6.2 INCREASING COMPLEXITY OF PRICE CAP REGULATION

Price Caps with ARAMS-X. This is essentially the same as the preceding scheme, except that instead of using a general indicator of economy-wide inflation, an index which more closely reflects price inflation in the industry is substituted. This is designed to reduce the problems associated with errant indices. As outlined in Section 5.2, there are three main types of ARAMS.

Cost-Based Formula Pricing. This type of ARAMS re-establishes the link between the firm's costs and prices. Since the link is automatic, it generally offers even less incentive for cost minimization than does COS regulation and a greater incentive for strategic manipulation of inputs to increase profits. The main instrument for decreasing costs is increasing the value of the *X* factor. However, instead of choosing a value for *X* based on estimates of productivity growth, in this case it might be based on subjective judgements by the regulator as to how much costs could be reduced without jeopardizing the viability of the firm.

Restricted Own Cost ARAMS. These allow changes in specific costs incurred by the firm to be passed along in the form of higher prices. The two problems associated with this approach are that the firm has very little incentive to aggressively control the specific costs subject to the formula adjustment and it also provides an incentive for the firm to bias its choice of inputs in favour of those which have an automatic cost pass-through. For example, an automatic rate of return adjustment clause suggests an incentive for over-capitalization.

Cost Index Based ARAMS. These reduce the errant index problem and, at the same time, attempt to prevent a re-establishment of the link between prices and the firm's own costs. In the context of Canadian pipelines, using industry averages may not work given that these averages are essentially endogenous for large systems like those of NOVA and TCPL. Any use of an industry-wide cost index would largely provide the dominant companies with automatic cost pass-through of their own costs, effectively eliminating the efficiency advantages of the ARAMS. Furthermore, the fact that the weights in the index are based on the share in costs provides the firm with an incentive to strategically adjust its inputs in order to maximize profits.

Price Caps with Sliding Scale Plans. The performance of this version of price cap regulation will likely differ from RPI-X in the following ways:

(i) More of the cost reductions are passed on to customers since they benefit not only from the productivity offset, but also from an explicit share of the profit gains accruing to the firm.

(ii) For precisely the same reason as in (i), they provide less incentive for cost reduction.

(iii) Since sliding scale plans involve calculating rates of return, they are unlikely to have any information, accounting or monitoring advantages over COS regulation. However, since they control profits, they may enjoy more general acceptability.

(iv) To the extent that a major part of the cost reductions is passed along to consumers by lowering prices from what they would otherwise be, there is a reduction of the problem of a divergence of costs and prices. It will be recalled that this is an important concern with simple price caps.

Yardstick Competition. The practical difficulty which makes application of a yardstick scheme problematic is that it requires a large set of comparable firms to create a yardstick. 'Comparable' means that the firms have similar demand and cost functions. Moreover, they need to have similar investment histories, so that their rate bases and depreciation charges are comparable.

Preliminary work reported in Joskow and Schmalensee (1986, 35) shows that the relative rankings of electric utilities in the United States based on estimated costs is extremely volatile. They conclude that "... comprehensive yardstick approaches to rate setting would impose highly random rewards and punishments; inefficient utilities might prosper while efficient producers might not be viable, and prices would often be out of line with both actual and minimum attainable costs." Creating a meaningful yardstick for the major gas pipelines in Canada is likely problematic.

6.3 EVALUATION OF PROFIT LEVEL INCENTIVE SCHEMES

It might be recalled that these approaches generally attempt to provide greater incentives for the regulated firm to increase cost efficiency through changes in the allowed rate of return. In most cases they would be used in conjunction with COS regulation and the evaluation presented below assumes that they are combined with this type of regulation.

Cost of Service with Banded Rates of Return. The advantages of this scheme over traditional COS regulation are: (i) it reduces the regulatory burden associated with regulation since the number of rate hearings should be reduced, and (ii) the firm is provided with incentives to cost-minimize since any cost savings it achieves can be kept, at least until the upper limit or band for the rate of return is reached.

The two disadvantages of this scheme in the context of pipelines are:

(i) The transmission companies are put at somewhat greater risk (although this risk is still limited by the lower limit of the deadband) which could potentially increase the cost of capital and, consequently, the level of tolls.

(ii) Assuming a rate hearing is triggered when the actual rate of return moves outside of the deadband, the incentives of the firm to produce efficiently can become perverse. For example, once the actual rate of return reaches the upper band, any additional incentives to increase efficiency disappear and are replaced by incentives to reduce efforts to cost-minimize. These perverse incentives arise if the firm anticipates that triggering a rate hearing will result in new tolls or rates which reflect actual costs. In such a case, the firm has an incentive to keep its rate of return just below the upper limit, perhaps by engaging in pure waste or gold-plating. Similar incentives exist for it when its rate of return ends up near the lower band. In this case it has an incentive to trigger a rate hearing as soon as possible and the incentives for cost efficiency are distorted by its incentives to build as strong a case as possible for higher rates.

Cost of Service Regulation with Profit Sharing. This scheme also creates higher-powered incentives for the firm to minimize costs. With this approach, the firm is supposed to be able to keep a permanent share of the benefits from cost minimization. The main disadvantage is that the incentives for cost minimization depend on the ability of the regulator to commit in advance that it will not, at a future rate hearing, set prices equal to actual costs. This kind of open-ended commitment is difficult for a regulator to make if, as is quite likely, customers view higher profits as a signal that regulation has failed rather than a signal that it has succeeded in generating savings. Such commitments are also problematic in that regulators usually cannot bind their successors. This system also reduces the share of benefits from cost reductions which go to customers.

Cost of Service Regulation with Benchmarking. The essential problem associated with introducing performance indicators or benchmarking is that it is difficult to relate the firm's achievements in specific areas to a measure of overall performance on which the rate of return could be based.

The use of performance measures and targets other than profitability may be appropriate in the management of individual departments, in designing facilities or in the ongoing monitoring of the overall operations of an individual company. However, there are major problems when these are incorporated as a component of the formal regulatory process for the purpose of determining allowed costs and rates of return.

Measurement Problems. The basic measurement problem is in

deciding what is good performance. Judgements regarding the separation of good performance from bad performance require a measurement standard. This poses the vexing problem of what standard to use. In the absence of information about cost and demand functions, the possibilities are intra-industry, inter-industry, and intra-industry / inter-country comparisons. There is not a uniform or consistent life-cycle pattern for pipelines and there are large differences across Canadian pipeline systems (let alone pipeline systems in other countries) in terms of age, configuration, overall size and other factors. This can make inter-pipeline comparisons virtually meaningless in many cases.

There are numerous variables such as quality and reliability that are inherently difficult to define and measure in the case of pipelines. Further, even if agreement is reached on the appropriate measure for purposes of pipeline design and operation, this measure would not be appropriate for evaluating actual performance. For example, consider the case of reliability and *assume* that there is agreement that the appropriate design standard is a Loss of Load Probability equal to y. This is an *ex ante*, probabilistic measure which gives expected reliability.

However, such a measure would be irrelevant in terms of evaluating actual reliability performance for the purpose of establishing allowed costs or the allowed return. Rather, for this purpose it would presumably be a measure such as the observed number of reliability incidents over some recent historical period. Since facility failures are typically random, the number of actual failures in any year could be quite different from the expected frequency used in the approved design of the system. Consequently, tying allowed costs and/or returns to the number of observed failures would unfairly reward the company in those years where there were fewer-than-expected reliability incidents and unfairly punish it in those years where there were more than expected. Over an extended period of time, the number of actual incidents would converge toward the expected number upon which the facilities were designed. However, the reward/punishment in any period would be random and not reflect the quality of management.

Index Problems. As noted in the CAPP submission to the NEB's Workshop On Incentive Regulation (CAPP 1992), there are many aspects of specific performance measures such as 'quality of service' and many individual performance measures (for example, operating and maintenance costs per unit of throughput, employees per unit of throughput, etc.). In order to be useful in judging performance in a regulatory context, these must be reduced to a single measure or index consisting of a weighted average of the individual measures.

However, it is not possible to construct a meaningful index since detailed information about the cost or production functions is not generally available. As a result, there is no way of determining unequivocally whether overall performance went up or down, or was high or low, based on the pattern of changes in a set of ad hoc measures. For example, it is impossible to say anything at all about efficiency or the performance of management for a combination such as a rise in the number of employees per unit of throughput, a decline in the rate base and an increase in operating and expenditures per unit of throughput. Such a configuration could be consistent with exceptionally good or exceptionally poor performance by management.

Tradeoffs, Distortions and Biases. Experience with performance measures and targets in other regulatory jurisdictions demonstrates the likelihood of a greater distortion of incentives and reduced efficiency. The problem with focusing on specific aspects of the operation of a firm is that this creates adverse incentives which favour those aspects over others. The firm will obviously meet the required standard(s), but usually at a cost to overall performance. If the target is, say, reduced unit operating and maintenance costs, this becomes the managerial focus even though it might mean reduced overall efficiency and higher overall costs. For example, such a focus may lead to earlier replacement of equipment than would otherwise be the case.

Berg and Jeong (1991) provide a systematic assessment of the overall effect of these types of programs for electric utilities in the United States. They conclude that such programs did not significantly improve the overall cost performance of the firm. They speculate that these programs may be ineffective because "... by focussing on specific categories or determinants of cost, regulators induce utilities to devote excessive resources to ensuring that a narrow goal is reached—so no net cost savings are realized."(Berg and Jeong 1991, 53)

Regulatory Burden. Rather than a more streamlined and less adversarial regulatory process, performance standards and targets in setting allowed costs and returns can often lead to the opposite result. For all of the reasons summarized above, they invite arguments about definitions, measurement and meaningfulness. They introduce additional subjectivity and, contrary to the intent of most incentive schemes, result in greater regulatory burden and more intervention by regulators in the management of regulated utilities.

Cost of Service Regulation with Capital Cost Incentives. In the context of the large gas pipelines in Canada, initiatives which control construction costs have the potential to significantly lower the cost of service since the recovery of such costs represents a major part of total costs.

The main disadvantage of these schemes is that they involve considerable up-front administrative costs and expertise on the part of the regulator to institute effectively. Their effectiveness depends on the regulator's estimation of the construction costs. If the regulator has relatively good information regarding the potential costs, it need not provide the firm with as large an incentive to truthfully reveal its best estimate. Moreover, it may run into acceptance difficulties based on its underlying complexity and the notion that firms should not be rewarded for doing something they should do anyhow.

While alternatives such as expedited certificates may avoid these excessive information requirements, they are subject to other problems. For example, they create considerable uncertainty as to whether an expansion will eventually be tolled on a rolled-in or an incremental basis. This uncertainty implies greater risk which could significantly affect the cost of financial capital and, in turn, the overall level of tolls. Further, as discussed in Section 4.3, incremental tolling often has significant negative consequences in terms of proper price signals and overall efficiency. If incremental tolling is selected after review of facilities constructed under expedited certificates, the inefficiency which they generate could be much larger than any efficiency gains via reduced capital costs.

6.4 REGULATORY ALTERNATIVES IN PRACTICE

The purpose here is to highlight some of the important practical aspects of the two main approaches to regulation in the context of major gas pipelines. The focus is on the price cap version of price level regulation. It will be recalled that a number of the other alternatives set out in Section 5 (such as Price Caps with ARAMS, with Sliding Scale Plans, and Yardstick Competition) are all versions of regulatory schemes involving caps on prices that can be charged by the firm.

Sustainability and Applicability of PCs to Pipeline Regulation. One of the conclusions which follows from Section 6.2 is that price regulation applied to a natural gas pipeline, such as that under price caps, will likely eventually regress back to COS regulation. One reason is that price cap regulation was not designed as a regulatory regime for industries where the technology is mature and the recovery of large, sunk capital expenditures is important. In the words of the original proponent of price cap regulation for British Telecom, Steven Littlechild:

> The purpose of such a constraint is to reassure customers of monopoly services that their situation will not get worse under privatization. It 'holds the fort' until competition arrives, and is inappropriate if competition is not expected to emerge. It is a temporary safeguard, not a permanent method of control. The one-off nature of the restriction is precisely what preserves the firm's incentive to be efficient, because the firm keeps any gains beyond the specified level. Repeated 'cost-plus' audits would destroy this incentive and, moreover, encourage 'nannyish' attitudes towards the industry. (Beesely and Littlechild 1986, 41)

In an assessment of price cap regulation in the United Kingdom, Beesely and Littlechild (1989) conclude that it is inappropriate for natural gas transmission and any application to that industry would be almost indistinguishable from COS regulation. Their view is that a necessary condition for applicability of price cap regulation is rapid technological change which fosters or allows competitive entry. The technological change must not only be rapid, it must change the cost minimizing industry structure. The technological change which has occurred in long-distance telecommunications means that it is no longer a natural monopoly. Therefore, the invisible hand of competition can replace the visible hand of the regulator to control market power.

Such is not the case in pipelines. As discussed in Section 3, the regulation of the major pipeline systems in Canada is typically justified on economic grounds. Significant scale, network and other economies make workable competition unattainable. Further, regulation is required to deal with the special requirements of an industry where there are large, sunk capital investments and where the capital has a very long economic life.

This issue aside, the experience in the U.K., where price caps have been in effect for a considerable period of time, would suggest that there may be a strong tendency for new-style regulation to essentially become traditional COS regulation with a different name. For example, as noted in The Economist (1995, 74), there have been frequent adjustments to the cap in order to curtail profitability of the regulated firm and reduce consumer prices. This, combined with the resulting swings in the share prices of the utility and the cost of financial capital, may in fact lead to greater distortions in investment patterns and efficiency incentives than under COS regulation. At the very least, this experience indicates the difficulty of achieving the regulatory commitment necessary to create strong and lasting efficiency incentives under price cap regulation.

In summary, unless these types of difficulties can be adequately addressed it is far from clear that price cap regulation for the major gas pipelines would be superior (in terms of the key regulatory criteria) to traditional or streamlined COS alternatives.

Essential Differences Between COS and PC Regulation in Practice. The two main differences between COS regulation and price cap regulation are:

(i) Price caps substitute a formula and periodic reviews for rate hearings to adjust prices to costs as costs and demands change over time.

(ii) Under price caps the firm does not require regulatory approval to expand capacity.

To the extent that the substitution of a formula for rate hearings increases risks due to the 'Russian roulette' problem associated with errant indices, there will be demands on the regulator to have more frequent rate hearings to true-up costs and prices. Recall that capital costs are the major cost item for the major gas pipelines and, consequently, variations in factors such as interest rate levels will have a major effect on overall costs.

It is unlikely that the overall price index used under price cap regulation would capture these variations. In fact, if such a regime had been in place when interest rates and the cost of equity capital were rising, tolls would have had to follow a downward trend (given low rates of inflation, combined with the productivity or *X* factor) even though the cost of capital would be increasing. In circumstances like these there would be demands to re-set or re-adjust prices and any failure by the regulator to meet these demands would mean that the viability of the firm becomes threatened. In the absence of these readjustments, the result may be to substantially increase the risk for both the pipeline and shippers, leading to inefficient levels of investment by both. As noted in Section 5.1 and in the discussion of the price cap scheme recently applied to IPL (see Section 6.1), it is important to incorporate off-ramps to allow rates to be adjusted under such circumstances.

Related to the 'Russian roulette' problem is the issue of optimal capacity levels. As indicated in Section 2, it is likely that significant expansions to the major gas pipeline systems will be warranted over the foreseeable future. These expansions will be key determinants of the level of overall costs and tolls. However, such factors do not typically enter into the price-setting formulas embodied in price-level regulation schemes like price caps. This means that either the regulator would have to re-open the issue of the appropriate level of tolls to accommodate the expansions or the expansions would not materialize. If a hearing or less formal procedure (such as a negotiated settlement) were used to re-set tolls, there would really be no gain over the more traditional COS schemes. If they were not re-set, the consequences in terms of lost opportunities and the associated costs for the natural gas industry would be negative.

It might be recalled that the price cap scheme recently adopted for the regulation of IPL includes a provision for adjustments in the case of significant capacity additions or other changes in service levels.[108] However, other price cap proposals for Canadian gas pipelines do not incorporate these types of off-ramps.[109] The implicit assumption is that these pipelines would still undertake the expansions but this is unrealistic unless a significant readjustment to its tolls was allowed. It might also be noted that if the regulator uses profit data to set rates, given the likely frequency of rate hearings arising from the errant indices problem and capacity expansions, price cap regulation essentially regresses back to COS regulation.

Other Implications. A substantial departure from traditional regulatory approaches can have fundamental, and possibly irreversible, implications for the major gas pipelines. Of particular importance would be the implications related to impacts on the degree of regulatory risk, the degree of exposure to other risks, the likely range for expected rates of return, and the potential for expansion.

The Level of Risk. There are several types of risk relevant to any evaluation of regulatory regimes from a private or corporate perspective. One is regulatory risk. In general, this concerns the potential for significant changes over time in the nature of regulators' decisions (operating under a given regulatory regime) to adversely impact the expected earnings of the firm. The other main type is the risk inherent in each of the various regulatory regimes. For example, under traditional COS regulation there has typically been a 'regulatory compact' in effect. This has, among other things, involved an implicit understanding that the regulated firm would be protected from most risks arising from factors beyond its control or which could not be reasonably foreseen. On the other hand, regimes such as price cap regulation can involve the firm accepting significantly more risk in return for somewhat greater flexibility with respect to pricing, input and expansion decisions.

---

108 See Section 6.1.
109 See Imperial Oil Limited (1994).

A key issue is how financial markets would perceive any such changes in these two types of risk because of a shift in regulatory regimes. The result could be a significant increase in the risk premium component of the cost of financial capital, particularly if other comparable, regulated firms in other jurisdictions continued to be regulated using the more traditional approaches.

This issue is particularly important for two reasons. First, capital costs (including capital recovery) represent the major cost item for the major gas pipelines. Any additional risk premium in the cost of funds could, therefore, have quite a significant effect on tolls. This, in turn, may negatively impact long-run earnings potential by affecting the amount of future expansion of facilities. Second, it emphasizes that what is really of importance is not the level of the return but rather the level of the risk-adjusted rate of return. To use an example, one can observe that some of the oil pipelines subject to light-handed regulation have earned higher returns in recent years than the Group 1 gas pipelines. An interesting issue is whether this differential reflects a difference in real performance or a difference in risk.

The relationship between risk and the optimal (financial) structure is also relevant. If, for instance, a switch to a new regulatory regime can be shown to significantly increase the risk facing the company, it may be necessary to increase the equity ratio. Alternatively, if the change put the firm at greater risk but there were no increases in the allowed equity ratio, the firm's long-term profitability would be threatened. The equity ratio has important ramifications for the overall revenue requirement and the level of tolls. It has been observed, for example, that equity ratios tend to be higher for U.S. pipelines than for those operating in Canada. This generally reflects the greater degree of risk facing shareholders under the U.S. regulatory system. The conclusion is that, in addition to raising the cost of financial capital by increasing the risk premium, any significant rise in the level of risk will have the tendency to increase total capital costs by raising the equity ratio. Both effects work to increase tolls from what they would otherwise be.

Short-Term vs. Long-Term Gains. An important factor in any evaluation of the various alternatives is the appropriate corporate discount rate. As demonstrated earlier, most of the incentive alternatives hold out the possibility of higher rates of return for at least a short period of time. Then, either a review is held to re-set the rates / tolls or some formula is used to reallocate the extra earnings. It is highly probable under some of the regimes that, in practice, the higher the earnings, the higher the hurdle that will be set after the review or re-setting. Given this, it is clear that many schemes will involve a potential short-run gain in earnings but possibly lower longer-term rates of return. Such alternatives may only be attractive to regulated firms which employ above-average discount rates.

A related consideration has to do with the firm's estimate of potential cost reductions. If it believes that costs can be significantly reduced, the alternatives which tie the rate of return to cost performance become attractive, at least in terms of short-run gains. However, once the 'easy' cost reductions have been made, it may be much more difficult to have such a favorable cost savings record in the future. If this is the case, the initial earnings gain cannot be sustained and, under some schemes, the likely result may be lower future earnings.

7 Summary and Conclusions

7.1 STUDY OBJECTIVES

There have been dramatic changes in natural gas markets in the past decade. The deregulation of gas markets and prices in both the U.S. and Canada has served to increase the intensity of competition. In addition, with the supply overhang from the earlier era, the decline in real prices, and the need to reach markets further away to absorb the growth in production capacity, there has been a sharp focus by the industry in Canada on efficiency issues.

Since gas transmission costs represent a large portion of the total delivered price of natural gas in many markets, it is not surprising that these events would be accompanied by pressures to reduce gas transmission tolls. This has led to requests by shippers for changes in the regulatory regime to generate stronger incentives for pipelines to become more efficient. Further, these requests have often used the dramatic changes in regulatory regimes in areas like telecommunications as evidence that alternative approaches to regulation are viable and can lead to significantly lower tolls.

A number of developments offer the potential for further progress in defining and evaluating alternative regulatory regimes. Along with the rapid growth in the literature on regulation, various hearings and workshops have significantly augmented knowledge concerning alternative regulatory regimes. In addition, there has been considerable experience and knowledge accumulated in the recent past from regulatory changes and experiments for a variety of utilities and in a number of different jurisdictions.

The main objective in this study was to bring together and apply these developments in an examination of the main alternative schemes for the regulation of the major natural gas pipelines in Canada. A particular goal was to provide a framework that would be useful in clarifying and analyzing some of the key issues. A further objective was to present concepts, techniques and research results that might advance the debates on the regulatory alternatives and possibly assist in resolving some of the important issues.

The main regulatory alternatives considered include traditional cost of service (COS) regulation, streamlined COS regulation, deregulation and market approaches, price level incentive regulation (such as price caps, automatic rate adjustment mechanisms, sliding scale plans, and yardstick competition), and incentive schemes that operate via changes in allowed profits (such as banded rates of return, profit sharing, benchmarking and capital cost incentives) and are usually combined with some form of COS regulation.

7.2 ROLE AND CHARACTERISTICS OF GAS TRANSMISSION

Section 2 of the study provides an overview of the development and structure of the gas pipeline industry, and a discussion of the changes in the market and regulatory environment that have important implications for the regulation of pipelines. The main points are summarized below.

(i) The future increases in Canadian transmission and productive capacity are unlikely to be as large as they have been since the deregulation of gas markets and prices in the mid-1980s. However, significant growth over the foreseeable future is expected. This will require penetration of export markets which are further away and highly competitive. One consequence will be a strong focus on transmission costs but, at the same time, recognition of the need to accommodate the costs associated with further required expansion of capacity.

(ii) The rapid growth in capacity for the major gas transmission systems has meant that capital costs (return on rate base and capital recovery / depreciation) represent a major portion of total costs. With continued expansion, this characteristic will remain. The proportion of total costs accounted for by operating and maintenance expenditures is typically much smaller than the proportion represented by capital costs. As a result, it can be argued that there are potentially greater benefits to regulatory alternatives which focus on minimizing the costs associated with capital than those which have a primary goal or objective of reducing total operating and maintenance costs.

(iii) The greater reliance on market forces in natural gas pricing means that load factors are likely to remain higher and show less variability than was the case prior to the mid-1980s. Also, with the loss of the merchant function, there is limited ability by pipeline management to affect the size and directions of gas flows and there is limited ability to rely on pipeline competition in Canada to constrain tolls or managerial behaviour. Further, the issue of reliability and quality of service is very important.

(iv) In general, the Canadian regulatory regime, or changes thereto, should take into account that: the main costs are those associated with the financing and recovery of the large amounts of capital involved in gas transmission; investment decisions concerning the timing and size of capacity are a major determinant of tolls and usually much more important than operating decisions once the capacity is in place; toll patterns over time are more a function of exogenous factors such as depreciation, capacity expansions and interest rates than they are a function of factors that can be controlled by pipeline management such as the rate of adoption of new technologies or labour productivity; increments to Canadian capacity are frequent and large, and the resulting toll adjustments are not amenable to 'simple formulas'; and, there remain valid reasons for continued regulation of tolls and other aspects of the major gas transmission systems in Canada (these reasons include the existence of certain natural monopoly elements and the classic hold-up problem associated with incomplete contracts and the large, sunk and irreversible aspects of investments).

(v) While some of the arguments for changes in the regulation of pipelines are based on the belief that tolls could be significantly reduced if the appropriate incentives for efficiency were put in place, others are often related to the observed regulatory revolutions in areas such as telecommunications. However, it is noted that: the arguments that transportation costs can be reduced markedly, just as the costs of the producing sector have been reduced by large amounts, have questionable validity; the dramatic regulatory changes in the case of other utilities such as telecommunications and electricity reflect to a large degree substantial technological changes which allowed competition to replace natural monopoly; there have been no such changes in the case of natural gas transmission; technological change in the case of gas pipelines has tended to increase rather than decrease the economies of scale and scope that lead to natural monopoly and this reduces the possibility of workable competition as an alternative to regulation; and the move to alternatives such as price cap regulation in the case of other utilities often entails circumstances quite different from those in the case of major Canadian gas pipelines.

7.3 REGULATORY CONCEPTS AND OBJECTIVES

Section 3 outlines the main theories of regulation, the rationale for regulating natural gas pipelines, the rules for optimal pricing and the standard criteria used for evaluating regulatory regimes and tolling methodology.

(i) Although there are other theories of regulation, such as those which view regulation as a means of redistributing wealth in response to special interests, the *Public Interest Theory* of regulation is viewed as the most appropriate in the context of pipeline regulation.

(ii) Under this general theory, regulation is justified when there is a significant case of 'market failure' and where it can be shown that the social costs of not intervening are materially greater than the costs associated with imperfect regulation. The main sources of market failure in the case of the major gas pipelines in Canada relate to network economies, natural monopoly elements and the large, durable and sunk nature of the investments. Given these, transmission costs may be greater if there were competing systems, any competition achieved may not be sustainable over the longer term, and the amount of investment in transmission and productive capacity could be sub-optimal under a competitive alternative. The economic objective of regulation in such a case is to intervene so that the firm providing transportation service sets prices and levels of service as close as possible to those consistent with efficient outcomes.

Long-term contracts are often required to underpin investments in transmission capacity. However, such contracts are necessarily incomplete in that they cannot incorporate all contingencies. In this context, an important role of the regulator is to allow efficient adjustments to contingencies or unforeseeable events not covered by private contracts.

(iii) There is evidence that competition at the city gate, along with the potential of new entrants, is likely sufficient to eliminate the potential for significant monopoly pricing by pipelines serving some U.S. markets in the event that tolls were deregulated. For pipelines serving other U.S. markets, and for the major ex-Alberta pipelines serving domestic and export markets, it would appear that such deregulation would allow the exercise of significant market power in some circumstances. Further, it is not clear that additional competition in these circumstances is socially desirable. Perhaps the most important competition which exists in Canada is the competition for the next expansion. However, there are also a number of limitations with this type of competition.

(iv) In general, it is not possible to have 'first-best' prices, where services are priced at their marginal cost, without violating the revenue constraint (that revenues collected just cover the actual costs) of a regulated firm. That is, in practice, prices will deviate from marginal cost and 'optimal pricing' will concern the setting of prices so as to minimize the distortions and inefficiencies associated with deviations from marginal cost. Ramsey pricing is the benchmark used for assessing the efficiency aspects of prices when there is a binding regulatory constraint. It is also noted that two-part tariffs (where there is a fixed access or demand charge and a per-unit or commodity charge) are beneficial in terms of reducing the distortions associated with deviations between prices or tolls and marginal cost.

(v) The set of standard criteria used for evaluating the various regulatory alternatives include: efficiency (minimizing costs and setting prices that give the right market signals); consumers should benefit from cost reductions; viability / sustainability and stability (the regulated firm should remain viable, the regulatory regime should be sustainable and the regime should promote rate stability); fairness / equity (for example, tolls should be 'just and reasonable,' there should be 'no unjust discrimination' and the process should be perceived as fair); regulatory burden (the costs incurred should not be onerous); and implementability (the regime should embody practicality, administrative simplicity and general acceptance).

7.4 TRADITIONAL REGULATORY REGIMES

Section 4 focuses on traditional cost of service (COS) regulation. Along with a description of the key elements of the various forms of COS regulation, it provides a discussion of the resulting toll patterns and an evaluation of its overall performance in terms of the standard regulatory criteria.

(i) The approach that has been used by the NEB to regulate Group 1 gas pipelines generally corresponds to traditional COS regulation. In practice, it often entails considerable regulatory oversight with regard to costs and tolls, in addition to regulation of the rate of return. Key points with respect to this type of regulation are: it typically involves a 'regulatory compact' that minimizes the risks to the firm; there is some flexibility with respect to tolling patterns/levels under such regimes (for example, although front-end loading is common, it is possible to move to more levelized patterns to improve the competitiveness of tolls); there are other types of pricing flexibility that can be incorporated such as through the use of bands in relating tolls to allocated costs and the use of multipart tolls with different allocations of fixed and variable costs; further variations involve the use of *ex post* rather than prospective (or test-year) ratemaking and the use of light-handed approaches similar to those employed by the NEB for Group 2 pipelines or those that have been employed by the Alberta Public Utilities Board in regulating the NOVA gas transmission system.

(ii) COS regulation with streamlining differs from the traditional approach in that it incorporates initiatives to reduce regulatory burden, to improve the responsiveness of the firm to its customers and, in general, to use less adversarial approaches to resolve issues. This streamlining could include the extensive use of such things as: task forces and committees; negotiated settlements; generic hearings or formulas to set allowed rates of return; and exemptions from formal hearings for routine investments and those previously considered as part of a multi-year plan. A number of these types of streamlining have recently been adopted by the NEB.

(iii) Traditional COS regulation performs quite well in terms of most of the standard 'public interest' criteria. Contrary to the views expressed by some critics, COS regulation, as usually applied, does incorporate some important cost reduction and efficiency incentives (an example is the incentives associated with regulatory lag if the scheme is applied on a 'test-year' basis and the lag is significant). By minimizing risk to the pipeline company, it has the effect of minimizing the cost of capital. That is, it results in low risk premiums for financial capital and the ability to finance the utility with a high debt–equity ratio. This has important ramifications for overall toll levels given that capital costs are often the dominant item in total costs for a major transmission system, particularly one which has been expanding.

Further, it does allow for the possibility of some degree of pricing flexibility (for example, through the use of alternative depreciation methodologies). The main shortcomings are that it can entail a fairly substantial regulatory burden and the incentives for cost minimization may be insufficient if regulatory auditing is inadequate in terms of frequency or rigor. The absence of a significant regulatory lag is also typically viewed as reducing the incentives for cost efficiency. Also, in some cases it can be argued that the incentive to minimize the costs associated with the installation of capacity is not particularly strong.

(iv) Streamlining can significantly increase the performance of COS regulation in terms of the main 'public interest' criteria. In particular, the use of such mechanisms as negotiated settlements and formula approaches to return on equity determinations can significantly reduce regulatory burden. However, there is also a possibility that streamlining can lead to the perception by some stakeholders that the procedures are less fair and the outcomes are less than equitable.

7.5 INCENTIVE ALTERNATIVES

This section begins by highlighting the important considerations in designing and implementing incentive regulation. Four incentive schemes which focus on setting the prices that can be charged are then outlined. These schemes are price caps, various automatic rate adjustment mechanisms, sliding scale plans, and yardstick competition. Next, the incentive alternatives that primarily operate through changes in allowed profit rates are summarized. These include profit sharing, banded rate of return, benchmarking and capital cost incentives. The final part of Section 5 discusses some other approaches such as deregulation, franchising, complete secondary markets for capacity, and various schemes to allow for greater pricing flexibility.

(i) The literature on 'optimal regulatory regimes' is lacking in terms of presenting practical or implementable schemes. However, along with the literature on incentive regulation, it does provide useful insights in designing and implementing effective incentive regimes. Some of the conclusions are that good incentive schemes: should promote the use of the regulated firm's superior microeconomic information; can only realistically achieve one or two objectives simultaneously and, hence, the regulatory objectives must be highly prioritized; should focus on the most sensitive and least random measures of performance; should only attempt to hold the firm accountable for outcomes which are clearly under the firm's control; should generally use broad rather than narrow-based performance measures; must pay particular attention to establishing appropriate performance benchmarks; should, if at all possible, entail several optional plans from which the firm must select one; should not promise more *ex ante* than the regulator can deliver *ex post*; and should clearly indicate the duration of the plan and the conditions under which it can be re-opened.

(ii) The incentive alternatives which focus on prices essentially fix the level of allowed prices by some formula and give the firm greater flexibility in all cost-related decisions.

Price Caps (PCs). Under this approach a weighted index of prices for the services provided by the firm is adjusted over time by the change in some economy-wide price index, less an *X* factor. This factor is supposed to reflect productivity growth. In some applications, the firm is allowed to adjust relative prices and to make decisions about capacity expansions without regulatory interference.

Automatic Rate Adjustment Mechanisms (ARAMS). These are similar to the PCs indicated above, except that the economy-wide inflation index is replaced with some other index. The three main variations involve an index that measures changes in the firm's input costs, an index that measures changes in only certain key input costs, or an index that measures changes in the average input costs for the industry.

Yardstick Competition (YC). Under this scheme, the tolls or prices for a particular regulated firm are set based on those observed for other comparable firms operating in a competitive environment or the regulated firm's prices are adjusted based on price changes for comparable firms operating in a competitive environment.

Sliding Scale Plans (SSPs). These are usually incorporated in PC schemes and involve adjustments to prices or tolls based on the realized rate of return. For example, if the rate of return achieved exceeded a predetermined level, some share would be returned to customers in the form of a rebate or reduced prices. If the actual rate of return fell below a predetermined level, a rate hearing would be triggered.

(iii) The incentive schemes which primarily operate by altering the level of profits typically represent an attempt to reward the firm for good cost performance by allowing it to increase its rate of return. In most cases, these schemes are applied in the context of COS regulation.

Banded Rate of Return (BROR). Under this approach, a band is set for the rate of return. So long as the rate of return realized by the firm is within this band there is no regulatory action. If it moves outside this band, tolls are adjusted, either through some sliding scale scheme or by holding a rate hearing.

Profit Sharing (PS). This is a variant of traditional profit level regulation. Under it, cost reductions or increased efficiencies are shared between customers and the shareholders of the firm. For example, if the firm was authorized to earn 10 percent and at the next hearing it was determined that the firm actually earned 15 percent, an amount equal to, say, one-half of the 5 percent excess earnings rate would be allocated to customers by adjusting prices downward, and the remainder would accrue to the shareholders of the regulated firm.

Benchmarking (BM). With this approach, the allowed rate of return depends on the firm realizing some pre-determined goals or standards as defined by performance indexes. The allowed rate of return is then set based on the firm's performance in terms of these standards.

Capital Cost Incentives (CCIs). These are mechanisms aimed at providing incentives for the firm to provide accurate estimates of the costs associated with capital projects and incentives to minimize these costs. Formulas are used to determine the size of the allowed change in the rate base relative to initial estimates and actual or realized investment costs. Other alternatives such as Expedited Certificates provide similar incentives to minimize construction costs.

(iv) Other incentive alternatives based primarily on market concepts are also considered.

Deregulation Based on Supply Basin Competition. Here, the approach involves reliance on competition among supply basins for the main markets to constrain any monopoly power of the pipeline. Another element may be the competition for the next expansion or new pipeline.

Franchising. Only a few variants of this approach have any potential for application to pipelines. One of these involves open-bidding for construction and ownership of new links, with the original pipeline firm remaining in control of operations.

Secondary Markets. This involves the development of integrated, competitive markets to allocate capacity rights. In addition to increasing rationing efficiency, the observed prices in these markets could provide valuable market signals for the purpose of determining optimal capacity expansions.

Flexible Pricing. These are essentially schemes aimed at achieving greater pricing efficiency by allowing the regulated firm more flexibility in adjusting relative prices (for example, by adjusting the price charges for service one type of service relative to the price charged for another type of service). Optional tariffs and priority pricing also represent interesting approaches.

7.6 EVALUATION OF INCENTIVE ALTERNATIVES

Section 6 provides a general assessment of the main incentive alternatives that focus on price and profit-level regulation. The first part provides a summary of various case studies involving incentive regulation. Following this is a general evaluation of price-level incentive schemes and then an assessment of profit-level incentive approaches.

(i) The case studies suggest that price caps, usually with some form of profit sharing, have been the most widely adopted approach to incentive regulation over the past decade or so. However, these have mainly been used in the regulation of telecommunications, an area where technological change has fundamentally altered the rationale for regulation. Other cases involve the use of price caps in the regulation of electricity, in setting rates of an integrated gas utility (British Gas), and in the regulation of oil pipelines.

In their most common usage, price caps have been adopted as a temporary measure until the industry becomes workably competitive. The record of price cap regulation in the case of electricity or oil pipelines is mixed. In some cases, the experience under the incentive scheme has been too limited to draw any firm conclusions.

However, it would appear that these schemes have not been particularly stable. In a number of instances the adjustments have been frequent and have served to undermine the incentives for greater efficiency. Other approaches, such as light-handed COS regulation or COS regulation with banded returns seem to have been much more stable.

(ii) The general conclusions with respect to the price-level incentive alternatives (price caps, price caps with ARAMS, price caps with sliding scale plans, and yardstick competition) include the following. The more commonly proposed RPI-X price caps can be effective in providing substantial incentives for the firm to reduce costs. However, this requires that the regulator can make the difficult commitment to not take into account the firms realized rate of return when re-setting *X* at the periodic reviews. Problems arise with price caps when they are applied in situations where there are likely to be requirements for significant capacity expansions. Either full examination of each expansion is still required, with tolls re-set at that time (in which case there is no real reduction in regulatory burden) or, if this examination / re-setting does not occur, there is unlikely to be any incentive for the firm to undertake the socially optimal expansions. The 'errant index' problem can be particularly acute and it will often require the shift to one of the versions, such as PC-ARAMS, where there can be greater confidence that the pricing index will track the firm's costs.

Some of the advantages of PCs may not be realized in applications to pipelines. For example, one of the potential advantages is that it may allow the firm greater flexibility in adjusting the relative prices of the services it provides. While the resulting price discrimination will increase profits and may improve overall efficiency, it is likely to come under attack from shippers on grounds that it violates equity or fairness criteria. Also, flexibility of the firm to decide on expansions outside of the formal regulatory process may not be very valuable if such a process is needed to re-set tolls so that the expansion is economic for the firm.

For many of these reasons, price caps have, in practice, often tended to regress back to traditional COS regulation. The addition of sliding-scale or profit-sharing plans may generally reduce the incentives for cost reductions but can increase the acceptability of the scheme when the regulated firm's returns increase significantly. Profit sharing may also be helpful in ensuring a closer correspondence between prices and costs over time. The use of yardstick competition regimes for the regulation of the major Canadian gas pipelines is likely to be problematic.

These various schemes can also have important implications for the regulated firm. In most cases they would be designed and implemented in such a way that the rate of return is likely to exceed that under traditional COS regulation, at least for a period of time. However, the experience in many of the case studies suggests that, at the first review, the *X* factor would likely be significantly increased or other adjustments would be made to bring down the rate of return that could be achieved. It may be reasonable to expect that these periodic adjustments would make it more and more difficult to achieve even the return that would be expected under COS regulation.

If the formula is set so that the price constraints are quite binding, there is a strong incentive for the firm to reduce operating and maintenance costs but also a strong incentive to avoid expanding capacity or service. On the other hand, if the formula is set so that the constraint provides some slack, there may be more of an incentive to meet requests for additional capacity but prices will be too high, resulting in allocative inefficiency. The scheme recently adopted for Interprovincial Pipe Line Inc. addresses the problem associated with expansions by allowing the formula to be re-set with each expansion. Similar off-ramps or adjustments are provided for other unpredictable events. In general, the greater the number of such provisions and the more frequently they are employed, the smaller will be the improvements in efficiency incentives over those under more traditional approaches.

Another potential concern is that the added risk to the firm under some of these schemes would raise the cost of capital. Given the importance of these costs in total costs in many cases, this would put considerable upward pressure on overall cost and toll levels. In general, these types of reasons suggest that the price-level incentive schemes discussed tend to be more appropriate in cases where the regulated activity is mature, where there are unlikely to be frequent or major capacity expansions, and where most costs are of the non-capital variety.

(iii) Many of the profit-level incentive schemes proposed would most realistically be applied within a COS framework. The main conclusions include the following. The *Banded Rate Of Return (BROR)* approach does offer some additional incentives for efficiency and can reduce regulatory burden. However, it also introduces some perverse incentives when rates of return reach the limits of the band. Also, the pipeline may be exposed to greater risk and this could negatively impact the cost of capital and toll levels.

Profit Sharing (PS) schemes also create stronger incentives to minimize costs. However, to be effective they require commitments from regulators that may not, at least in practice, be realistic in many cases. *Benchmarking (BM)* schemes are much more attractive at a superficial level of examination than when evaluated in detail. In practice, they pose serious measurement problems and can distort incentives and outcomes in many cases. Also, they can tend to result in greater regulatory burden rather than less. *Construction Cost Incentives (CCI)* have some attractive features, especially for activities such as gas transportation where the capital costs are a major component of total revenue requirements.

There are also a number of implications for the regulated firm. While most of these schemes can be designed and implemented such that they would provide some upside for the firm, it is questionable whether this is a likely outcome over the longer term. Experience might suggest that the possibility of significantly higher rates of return in the short-run will often involve a trade off in the form of lower rates of return in subsequent periods. An important implication of a shift to one of these approaches is that it involves at least the implicit acceptance by the pipeline that it will be exposed to greater risk than under traditional COS regulation. This additional risk may be limited but it nevertheless means abandonment of the traditional regulatory compact. A key issue is how this added risk will be perceived by investors and the extent of any increases in financial capital costs. This has significant implications for long-term cost and toll levels.

7.7 CONCLUDING OBSERVATIONS

The traditional COS approach represents a widely accepted approach to the regulation of many utilities where competition is not a viable alternative. In the case of major gas pipelines, this approach has been fairly durable. In part this is because it has proven consistent with certain key characteristics of these pipelines (such as their status as contract carriers, the frequency and lumpiness of capacity additions and the dominant role of capital costs), and changes have been incorporated over time to mitigate its worst features. For example, most recently there have been attempts to streamline COS regulation to reduce regulatory burden, increase flexibility and make it more effective.

The real issue now seems to be whether the attempts to improve the regulatory regime applied to these major gas transmission systems should be aimed at producing further evolution of COS regulation or more dramatic shifts to newer alternatives such as price caps.

After examining the many complex issues in the regulation of gas transmission and the various regulatory alternatives it seems apparent that none of the new-style incentive regimes represent a panacea. Nor is it possible to conclude that any one of these is uniformly better in terms of key evaluation criteria than the traditional COS approach, especially that which incorporates streamlining.

Rather, all regulatory approaches embody trade offs that must be considered in the context of the particular circumstances of each transmission system. In determining whether a specific alternative would be a significant improvement, several fundamental questions must be answered. The first is whether the regime is appropriate given the characteristics of the system. For example, some approaches are well-suited for stable, mature systems but deficient for those where frequent, large expansions will likely be required. Similarly, some are far better as short-term or transitional mechanisms than as regimes which must be sustainable. A second important issue is whether the trade-offs are acceptable. For instance, in many cases the alternative may result in greater cost efficiency and reduced regulatory burden, but at the expense of real or perceived fairness or at the expense of the overall level of tolls if the cost of financial capital or the equity ratio is increased. It will be important in every case to recognize all such trade offs and obtain a consensus on their acceptability.

It will also be important to consider combinations of the regimes surveyed in this study rather than each on a stand-alone basis. Given the particular circumstances of each system in terms of factors such as the degree of competition, the level of capacity utilization, and the rate of expansion, it will generally be preferable to tailor the regime to some extent to the particulars of the circumstances. It is hoped that, at the very least, the discussion of the many alternatives presented in this study will be helpful in this process. This survey, along with the summary of experiences in other industries and jurisdictions, can also hopefully be useful in identifying and avoiding the main problems and pitfalls in the design and implementation of new regulatory approaches.

References

Acton, J. P., and I. Vogelsang. 1989. "Introduction [to Symposium on Price-Cap Regulation]." *Rand Journal of Economics* 20: 369–372.

Alberta Energy. 1994. *Status of the Review of the Electric Energy Marketing Act (EEMA) and the Electric Industry Restructuring Initiative.* Edmonton: Government of Alberta.

Alberta Petroleum Marketing Commission. 1992. *Alternative to Traditional Cost Of Service Regulation.*

Alger, D., and M. Toman. 1990. "Market Based Regulation of Natural Gas Pipelines." *Journal of Regulatory Economics* 2: 263–280.

Averch, H., and L. L. Johnson. 1962. "Behavior of the Firm Under Regulatory Constraint." *American Economic Review* 52: 1052–69.

Baron, D. P. 1989. "Design of Regulatory Mechanisms and Institutions." In *Handbook of Industrial Organization*, Vol.2, ed. R. Schmalensee and R. Willig. Amsterdam: North Holland.

Baumol, W., and D. Bradford. 1970. "Optimal Departures from Marginal Cost Pricing." *American Economic Review* 60: 265–283.

Beesley, M. E., and S. C. Littlechild. 1986. "Privatization: Principles, Problems and Priorites." *Privatization and Regulation—The UK Experience.* Ed. J. Kay. Oxford: Claredon Press.

——. 1989. "The Regulation of Privatized Monopolies in the United Kingdom." *Rand Journal of Economics* 20: 454–472.

Berg, S., and J. Jeong. 1991. "An Evaluation of Incentive Rate Regulation for Electric Utilities." *Journal of Regulatory Economics* 3: 45–55.

Berg, S., and J. Tschirhart. 1988. *Natural Monopoly Regulation.* Cambridge: Cambridge University Press.

Bhattacharyya, S. K., and D. J. Laughhunn. 1987. "Price Cap Regulation: Can We Learn From the British Telecom Experience?" *Public Utilities Fortnightly* October 15: 22–29.

Bradley, I., and C. Price. 1988. "The Economic Regulation of Private Industries by Price Constraints." *Journal of Industrial Economics* 37: 99–105.

Braeutigam, R. R., and J. C. Panzar. 1989. "Diversification Incentives Under 'Price-Base' and 'Cost-Based' Regulation." *RAND Journal of Economics* 20: 373–391.

——. 1993. "Effects of the Change from Rate-of-Return to Price-Cap Regulation." *American Economic Review* 83: 191–198.

Brennan, T. 1989. "Regulating by Capping Prices." *Journal of Regulatory Economics* 1: 133–148.

Brenner, D. L. 1992. *Law and Regulation of Common Carriers in the Communications Industry.* Boulder, CO.: Westview Press.

Broadman, H. G 1987a. *Deregulating Entry and Access to Pipelines: Drawing the Line on Natural Gas Regulation.* Ed. J. Kalt and F. Schuller. New York: Quorum.

——. 1987b. "Competition in Natural Gas Pipeline Wellhead Supply Purchases." *Energy Journal* 8: 113–134.

Brown, L., M. Einhorn, and I. Vogelsang. 1989. *Incentive Regulation: A Research Report.* Washington, D.C.: Office of Economic Policy, Federal Regulatory Commission.

——. 1991. "Toward Improved and Practical Incentive Regulation." *Journal of Regulatory Economics* 3: 323–338.

Brown, S., and D. Sibley. 1986. *The Theory of Public Utility Pricing.* Cambridge: Cambridge University Press.

Burness, H. S., and R. H. Patrick. 1992. "Optimal Depreciation, Payments to Capital, and Natural Monopoly Regulation." *Journal of Regulatory Economics* 4: 35–50.

Canadian Association of Petroleum Producers. 1992. *Revitalizing Oil and Gas Pipeline Regulation.* Submission to the National Energy Board with respect to Incentive Regulation.

Canadian Gas Association. 1992. *Natural Gas in the Canadian Economy.* Don Mills, Ontario.

Church, J., and R. Ware. 1995. *Industrial Organization: A Strategic Approach.* Burr Ridge, Illinois: Richard D. Irwin, (forthcoming).

Coase, R. 1946. "The Marginal Cost Controversy." *Economica* 13: 169–189.

Cookenboo, L. J. 1955. "Production and Cost Functions for Oil Pipelines." *Crude Oil Pipelines and Competition in the Oil Industry.* Cambridge, MA.: Harvard University Press.

Davis, V. W. 1993. "Summary of the Status of Alternative Regulation in Telecommunications." *NRRI Quarterly Bulletin* 14: 161–66.

De Vany, A., and W. D. Walls. 1994. "Open Access and the Emergence of a Competitive Natural Gas Market." *Contemporary Policy Issues* 12: 77–96.

Demsetz, H. 1968. "Why Regulate Utilities." *Journal of Law and Economics* 11: 55–65.

Donlan, T. G. 1989. "Ringing Decision." *Barron's* March 13: 27–28.

Edison Electricity Institute. 1987. *Incentive Regulation in the Electric Utility Industry.*

Ellig, J., and M. Giberson. 1993. "Scale, Scope, and Regulation in the Texas Gas Transmission Industry." *Journal of Regulatory Economics* 5: 79–90.

Face, H. K. 1988. "The First Case Study in Telecommunications Social Contracts." *Public Utilities Fortnightly* April 28: 27–31.

Federal Communications Commission. 1989. *In the Matter of Policy and Rules Concerning Rates for Dominant Carriers.* Washington, D.C.

——. 1990. "Policy for Rules Concerning Rates for Dominant Carriers (Price Caps), Second Report and Order." *Federal Communications Commission Record* 5: 6786. Washington, D.C.

——. 1992. "Price Cap Performance Review for AT&T." *Federal Communications Commission Record* 7: 5322–5337. Washington, D.C.

——. 1993. "Revisions to Price Cap Rules for AT&T." *Federal Communications Commission Record* 8: 5205–5208. Washington, D.C.

Federal Energy Regulatory Commission. 1982. *Williams Pipe Line Company, Opinion No. 154.* Washington, D.C.

——. 1985. *Williams Pipe Line Company, Opinion No. 154-B.* Washington, D.C.

——. 1991a. *Buckeye Pipe Line Company, Opinion No. 360-A.* Washington, D.C.

——. 1991b. *Revisions to Regulations Governing Authorizations for Construction of Natural Gas Pipeline Facilities.* Washington, D.C.

——. 1994. *Re: Pricing Policy for New and Existing Facilities Constructed by Interstate Natural Gas Pipelines.* (Docket PL94-4-000, September). Washington, D. C.

——. 1995. *Pricing Policy For New and Existing Facilities Constructed by Interstate Natural Gas Pipelines, Policy Statement.* (Docket PL94-4-000, May 31). Washington, D.C.

Finsinger, J. and I. Vogelsang. 1985. "Strategic Management Behaviour Under Reward Structures in a Planned Economy." *Quarterly Journal of Economics* 100: 263–270.

Gallick, E. C. 1993. *Competition in the Natural Gas Pipeline Industry: An Economic Policy Analysis.* Westport, Conn.: Praeger.

Gilbert, R., and D. Newberry. 1994. "The Dynamic Efficiency of Regulatory Constitutions." *Rand Journal of Economics* 25: 538–554.

Goldberg, V. 1976. "Regulation and Administered Contracts." *Bell Journal of Economics* 7: 426–448.

Hansen, J. A. 1983. *U.S. Oil Pipeline Markets: Structure, Pricing and Public Policy.* Cambridge, MA: MIT Press.

Heal, D. W. 1990. "From Monopoly to Competition: Marketing Natural Gas in the UK." *Utilities Policy* 1: 54–64.

Hillman, J. J., and R. R. Braeutigam. 1989. "The Potential Benefits and Problems of Price Level Regulation: A More Hopeful Perspective." *Northwestern University Law Review* 84: 695–710.

——. 1989. *Price Level Regulation for Diversified Public Utilities.* Norwell, MA.: Kluwer Academic Publishers.

Hurst, C. 1992. "Liberalization and Regulation of Telecommunications." *Utilities Policy* 2: 13–24.

Imperial Oil Limited. 1994. *PRIDE: A New Vision of Pipeline Regulation* (draft).

Inside F.E.R.C. 1993a. "Oil-Pipeline Proposal is One-Way Street, Charge Shippers, Consumers". Washington: 1, 10–11.

——. 1993b. "Final Oil-Pipeline Rule Switches Index, Eases Protest Standards". Washington: 1, 8–9.

——. 1993c. "FERC's Oil-Pipeline Rule Won't Streamline Workload, Both Sides Warn". Washington: 1, 5–6.

Interprovincial Pipe Line Inc. 1995a. *Incentive Toll Proposal, Principles of Settlement, Effective April 1, 1995.* Calgary, Alberta.

——. 1995b. *In the Matter of an Application by Interprovincial Pipe Line Inc. for Orders Under Part IV of The Act, Approving a Negotiated Settlement Respecting an Incentive Toll Methodology and Associated Tolls and Tariffs.* Application to the National Energy Board (February). Calgary, Alberta.

Isaac, R. M. 1991. "Price Cap Regulation: A Case Study of Some Pitfalls of Implementation." *Journal of Regulatory Economics* 3: 193–210.

Jones, D. 1991. "Old Style and New Style Regulation of Electrics: The Incentive Connection." In *The Future of Incentive Regulation in the Electric Utility Industry.* Indianapolis, Indiana: School of Public and Environmental Affairs, Indiana University and PSI Energy.

Joskow, P. L. 1974. "Inflation and Environmental Concern: Structural Change in the Process of Public Utility Price Regulation." *Journal of Law and Economics* 17: 291–327.

Joskow, P., and R. Schmalensee. 1986. "Incentive Regulation for Electric Utilities." *Yale Journal on Regulation* 4: 1–49.

Kahn, A. E. 1988. *The Economics of Regulation: Principles and Institutions.* Cambridge, MA: MIT Press.

Klein, B., R. Crawford, and A. Alchain. 1978. "Vertical Integration Appropriable Rents and the Competitive Contracting Process." *Journal of Law and Economics* 21: 297–326.

Kwaczek, A., and P. Miles. 1995. "Are Pipeline Transportation Markets Becoming More Competitive?" Discussion Paper, Canadian Energy Research Institute. Calgary, Alberta.

Laffont, J.-J., and J. Tirole. 1993. *A Theory of Incentives in Procurement and Regulation.* Cambridge, MA: MIT Press.

——. 1986. "Using Cost Observations to Regulate Firms." *Journal of Political Economy* 94: 614–41.

Liston, C. 1993. "Price-Cap versus Rate-of-Return Regulation." *Journal of Regulatory Economics* 5: 25–48.

Loeb, M., and W. Magat. 1979. "A Decentralized Method for Utility Regulation." *Journal of Law and Economics* 22: 399–404.

London Times. 1988. "Industrial Gas Users Face Price Changes". London.: 23.

——. 1989. "British Gas Forces Prices Down". London: 27.

——. 1991a. "Consumers Win Boost From New Gas Prices". London: 21.

——. 1991b. "Gas Prices Curbed for Next Five Years". London: 1.

——. 1993. "Ofgas Revises its Evidence to MMC". London: 27.

——. 1994. "BT Ordered to Bring in Faster Price Cuts". London: 22.

Lyon, T. P. 1990. "Natural Gas Policy: The Unresolved Issues." *Energy Journal* 11: 23–49.

——. 1991. "Regulation with 20-20 Hindsight: Heads I Win, Tails You Lose?" *Rand Journal of Economics* 22: 581–595.

——. 1995. "Regulatory Hindsight Review and Innovation by Electric Utilities." *Journal of Regulatory Economics* 7: 233–254.

MacGregor, M. E., and A. Plourde. 1987. *Regulating and Deregulating Canadian Natural Gas.* A University of Toronto Energy Study, Report No. 87-3.

Mansell, R. 1994. *A Framework For Evaluating Tolling Methodology For Interstate Natural Gas Pipelines.* A Study Prepared for Foothills Pipe Lines Ltd. and Westcoast Energy Inc. and presented to the Federal Energy Regulatory Commission, Docket No. PL94-4-000. Calgary, Alberta.

Mansell, R., R. Wright, and W.Kerr. 1984. "An Economic Evaluation of Formula Pricing for Fluid Milk." *Canadian Journal of Agricultural Economics* 32: 3–24.

Milgrom, P., and J. Roberts. 1992. *Economics, Organization, and Management.* Englewood Cliffs, New Jersey: Prentice Hall.

National Energy Board. 1984. *Twenty-Five Years in the Public Interest.*

Ottawa: Supply and Services, Canada.
——. 1986a. *Reasons for Decision, TransCanada PipeLines Limited, Availability of Services.* Ottawa: National Energy Board.
——. 1986b. *Reasons for Decision, Trans Mountain Pipe Line Company Ltd.,* Ottawa: National Energy Board.
——. 1986c. *Annual Report.* Ottawa: National Energy Board.
——. 1987. *Reasons for Decision, Interprovincial Pipe Line Limited, RH-4-86.* Ottawa: National Energy Board.
——. 1989. *Reasons for Decision, Interprovincial Pipe Line Company, RHW-1-89.* Ottawa: National Energy Board.
——. 1990a. *Reasons For Decision, TransCanada PipeLines Limited, GH-5-89.* Ottawa: National Energy Board.
——. 1990b. *A Review of Toll Adjustment Procedures.* Ottawa: National Energy Board.
——. 1990c. *Annual Report.* Ottawa: National Energy Board.
——. 1991. *Annual Report.* Calgary: National Energy Board.
——. 1992a. *Reasons for Decision, Trans Mountain Pipe Line Company Ltd., RH-3-91.* Calgary: National Energy Board.
——. 1992b. *Reasons for Decision, TransCanada Pipelines Limited., RH-4-91.* Calgary: National Energy Board.
——. 1992c. *Annual Report.* Calgary: National Energy Board.
——. 1993a. *Incentive Regulation Workshop.* Calgary: National Energy Board.
——. 1993b. *Annual Report.* Calgary: National Energy Board.
——. 1994a. *Possible Changes to the Secondary Market for Natural Gas Transportation Services.* Calgary: National Energy Board.
——. 1994b. *Canadian Energy Supply and Demand, 1994 Report.* Calgary: National Energy Board.
——. 1995a. *Annual Report.* Calgary: National Energy Board.
——. 1995b. *Re: Application to the National Energy Board by Interprovincial Pipe Line Inc. for Orders Pursuant to Part IV of the National Energy Board Act, Approving a Negotiated Toll Settlement.* (March 24). Calgary: National Energy Board.
Navarro, P., B. C. Peterson, and T. R. Stauffer. 1981. "A Critical Comparison of Utility Type Ratemaking Methodologies in Oil Pipeline Regulation." *Bell Journal of Economics* 12: 392–412.
Neri, J.A., and K.E. Bernard. 1994. "Price Caps and Rate of Return: the British Experience." *Public Utilities Fortnightly* September 15: 34–36.
Neu, W. 1993. "Allocative Inefficiency Properties of Price-Caps." *Jour-*

nal of Regulatory Economics 5: 159–182.

Norris, J. E. 1990. "Price Caps: An Alternative Regulatory Framework for Telecommunications Carriers." *Public Utilities Fortnightly* January 18: 44–46.

O'Shea, D. 1993. "CARE Challenges Incentive Rate." *Telephony* April 19: 15–20.

Panzar, J., and R. Willig. 1977. "Free-Entry and the Sustainability of Natural Monopoly." *Bell Journal of Economics* 8:1–22.

Pawluk, C. 1995. "Subadditivity Tests of the TransCanada Pipeline System." Unpublished M.A. thesis, Department of Economics, The University of Calgary.

Petroleum Resources Communication Foundation. 1992a. *Our Petroleum Legacy.* Backgrounder Series, Calgary: PRCF.

——. 1992b. *Crude Oil.* Backgrounder Series, Calgary: PRCF.

——. 1992c. *Natural Gas.* Backgrounder Series, Calgary: PRCF.

Pierce, R. J. 1990. "Price Level Regulation Based On Inflation is Not an Attractive Alternative to Profit Level Regulation." *Northwestern University Law Review* 84: 665–694.

Plourde, A. 1986. *Oil and Gas in Canada: A Chronology of Important Developments, 1941–1986.* PEAP Energy Study No. 86-5, Institute for Policy Analysis: University of Toronto.

——. 1987. "On the Role and Status of Canadian Natural Gas Carriers Under Deregulation." *Journal of Energy and Development* 13: 1–25.

Price, C. 1992. "Regulation of the U.K. Gas Industry. The First Five Years." *Annals of Public and Cooperative Economy* 63: 189–206.

Public Utilities Fortnightly. 1988. "State Solicits Proposals for Telephone Rate Incentives." December 8: 46.

——. 1989a. "California Proposes Incentive Regulation for Local Telephone Exchange Carriers." September 28: 45.

——. 1989b. "California Approves Price Caps for Local Telephone Carriers." December 7: 65.

——. 1990a. "Refunds Ordered Under Telephone Incentive Regulation Plan." April 26: 42.

——. 1990b. "North Dakota Implements Telephone Price Cap Law." July 5: 49.

——. 1990c. "Washington Commission Rejects Incentive Regulation Plan." October 25: 92.

——. 1990d. "State Court Stops Commission's Telephone Deregulation Plan." November 22: 48.

——. 1990e. "Michigan Concludes Case on Telecommunications

Reregulation." December 6: 66.

——. 1991a. "Incentive Regulation of Local Exchange Telephone Carriers." July 1: 46–49.

——. 1991b. "Special Feature on Incentive Regulation." November 1: 49–60.

——. 1992. "Illinois Bell Incentive Rate Plan Eliminated." February 1: 36.

Regulatory Times. 1993. "Canadian Producers Challenge U.S. Oil Pipeline Rate Proposal." Calgary: 1–2.

Restrictive Trade Practices Commission. 1986. *Competition in the Canadian Petroleum Industry.* Ottawa: Supply and Services, Canada.

Riordan, M. H., and L. Cabral. 1989. "Incentives for Cost Reduction Under Price Cap Regulation." *Journal of Regulatory Economics* 1: 93–102.

Sappington, D. E. M. 1994. "Designing Incentive Regulation." *Review of Industrial Organization* 9: 245–272.

Sappington, D. E. M., and D. S. Sibley. 1988. "Regulating without Cost Information: The Incremental Surplus Subsidy Scheme." *International Economic Review* 29: 297–306.

——. 1992. "Strategic Nonlinear Pricing Under Price-Cap Regulation." *Rand Journal of Economics* 23: 1–19.

Schmalensee, R. 1979. *The Control of Natural Monopolies.* Lexington, MA: Lexington Books.

——. 1989a. "An Expository Note on Depreciation and Profitability under Rate-of-Return Regulation." *Journal of Regulatory Economics* 1: 293–98.

——. 1989b. "Good Regulatory Regimes." *Rand Journal of Economics* 20: 417–436.

——. 1989c. "The Potential of Incentive Regulation." *The Market for Energy.* Ed. D. Helm, J. Kay and D. Thompson. Oxford: Oxford University Press

Sharkey, W. 1982. *The Theory of Natural Monopoly.* Cambridge: Cambridge University Press.

Sibley, D. S. 1989. "Asymmetric Information, Incentives, and Price-Cap Regulation." *Rand Journal of Economics* 20: 392–404.

Spulber, D. 1989. *Regulation and Markets.* Cambridge, MA.: MIT Press.

Teece, D. J. 1986. "Assessing the Competition Faced by Oil Pipelines." *Contemporary Policy Issues* IV: 65–78.

——. 1990. "Structure and Organization of the Natural Gas Industry: Differences between the United States and the Federal Republic of

Germany and Implications for the Carrier Status of Pipelines." *Energy Journal* 11: 1–35.

Terzic, B., and J. McKinnon. 1988. "Gas in Britain: Regulation of a Privatized Former State Monopoly." *Public Utilities Fortnightly* May 26: 20–26.

The Economist. 1995. Economics Focus: "Most governments face the problem of how to regulate monopoly utilities. Britain thought it had the answer. Now it's not so sure." March 11: 74.

Train, K. 1991. *Optimal Regulation: The Economic Theory of Natural Monopoly.* Cambridge, MA.: MIT Press.

Veljanovski, C. 1993. *The Future of Industry Regulation in the UK.* London: European Policy Forum.

Vickers, J., and G. Yarrow. 1988. *Privatization: An Economic Analysis.* Cambridge, MA.: MIT Press.

Vogelsang, I. 1988. "Price-Cap Regulation of Telecommunications Services: A Long-Run Approach." In *Deregulation and Diversification of Utilities.* Ed. M.A. Crew. Boston: Kluwer Academic.

Vogelsang, I., and J. Finsinger. 1979. "A Regulatory Adjustment Process for Optimal Pricing by Multiproduct Monopoly Firms." *Bell Journal of Economics* 10: 157–71.

Wald, M. D. 1993. "Toll-free 800 Services Take a New Turn." *Networking World* May: 30–35.

Waterson, M. 1988. *Regulation of the Firm and Natural Monopoly.* Oxford: Basil Blackwell.

Williamson, O. 1976. "Franchise Bidding for Natural Monopolies—In General and With Respect to CATV." *Bell Journal of Economics* 7: 73–104.

——. 1985. *The Economic Institutions of Capitalism.* New York: Free Press.

Wilson, R. 1989. "Efficient and Competitive Rationing." *Econometrica* 57: 1–40.

——. 1993. *Non-Linear Pricing.* Oxford: Oxford University Press.